Fast and Fabulous!
Flavor Secrets

Fast and Fabulous!
Flavor Secrets

Judy Gilliard

THOMSON
DELMAR LEARNING

Australia Canada Mexico Singapore Spain United Kingdom United States

THOMSON
DELMAR LEARNING

Fast and Fabulous – Flavor Secrets
by Judy Gilliard

Vice President, Career Education Strategic Business Unit:
Dawn Gerrain

Director of Editorial:
Sherry Gomoll

Acquisitions Editor:
Matthew Hart

Developmental Editor:
Patricia Osborn

Director of Production:
Wendy A. Troeger

Production Editor:
Matthew J. Williams

Director of Marketing:
Wendy E. Mapstone

Channel Manager:
Kristin McNary

Marketing Coordinator:
David W. White

Cover Design:
Joe Villanova

COPYRIGHT © 2006 Thomson Delmar Learning, a part of The Thomson Corporation. Thomson, the Star logo, and Delmar Learning are trademarks used herein under license.

Printed in the United States
1 2 3 4 5 XXX 09 08 07 06 05

For more information contact Delmar Learning,
5 Maxwell Drive
Clifton Park, NY 12065-2919.

Or find us on the World Wide Web at
www.delmarlearning.com or
www.hospitality-tourism.delmar.com

ALL RIGHTS RESERVED. No part of this work covered by the copyright hereon may be reproduced or used in any form or by any means—graphic, electronic, or mechanical, including photocopying, recording, taping, Web distribution or information storage and retrieval systems—without written permission of the publisher.

For permission to use material from this text or product, contact us by
Tel   (800) 730-2214
Fax   (800) 730-2215
www.thomsonrights.com

Library of Congress Cataloging-in-Publication Data

Gilliard, Judy.
  Fast and fabulous! : flavor secrets / Judy Gilliard.
     p. cm.
  Includes bibliographical references and index.
  ISBN 1-4180-2997-1 (alk. paper)
  1. Quick and easy cookery. I. Title.
TX833.5.G5353 2005
641.5'55--dc22
                                                                    2005022874

NOTICE TO THE READER

Publisher does not warrant or guarantee any of the products described herein or perform any independent analysis in connection with any of the product information contained herein. Publisher does not assume, and expressly disclaims, any obligation to obtain and include information other than that provided to it by the manufacturer.

The reader is notified that this text is an educational tool, not a practice book. Since the law is in constant change, no rule or statement of law in this book should be relied upon for any service to any client. The reader should always refer to standard legal sources for the current rule or law. If legal advice or other expert assistance is required, the services of the appropriate professional should be sought.

The Publisher makes no representation or warranties of any kind, including but not limited to, the warranties of fitness for particular purpose or merchantability, nor are any such representations implied with respect to the material set forth herein, and the publisher takes no responsibility with respect to such material. The publisher shall not be liable for any special, consequential, or exemplary damages resulting, in whole or part, from the readers' use of, or reliance upon, this material.

To Jan and Harvey Izen
for their unwavering support and friendship

Contents

Acknowledgments viii

Introduction 1
   About Herbs and Spices 3
   Harvesting, Drying, and Storing Your
      Homegrown Herbs 6
   The Freshness Test 10
   Basic Herbs and Spices 11
   Herb and Spice Chart 35
   Extracts 38
   Herb Blends 39

Starters 41

Salads 59

Soups 77

Beans 85

Pasta 93

Seafood 113

Poultry 127

Meat 157

Vegetables 173

Potatoes, Orzo, Rice 189

Breakfast 201

Breads 209

Desserts 225

Index 251

Acknowledgments

- Jean Sheffield . . . who is always there to make sure my writings make sense, as well as being a true friend.
- Friends who are family: Liz, Ed, Kellyann, Kristen, Karen, and Erik Kazor; Ethan Herb; Alice Barlow; and Chris, Lori & Brittany Darrington, Amy and Dereck Knapp.
- Aunt Rose . . . whose kitchen started it all!
- Doug "Dougie" . . . my cousin who loves cooking and eating as much as I do. . . . Susan, Amanda, and Bryson Edgar.
- My sister, Teri . . . whose artwork truly amazes me.
- Cousins Chuck and Jean Gilliard.
- Uncle John and Aunt Ginny Hardacre.
- Matt Hart . . . who has made this all fun again!
- Patricia M. Osborn . . . for taking the time with me to make sure all is on order!
- Thomson Delmar Learning and their great staff.

Nutritional Analysis done with the software from Nutritional Data Resources.

Introduction

Fast and Fabulous—Flavor Secrets focuses on the modern active person who really doesn't have a lot of time to spend in the kitchen. Despite the lack of time, this person still wants to serve top-quality food and promote optimum health. For that reason, most of these recipes are simple, illustrating the basic use of herbs and spices.

I write cookbooks for the health conscious. If you equate "health conscious" with food that is neither tasty nor appealing, I want to change your thinking about that! In Flavor Secrets, I'll show you how easy it is to cook food that is fast to prepare and fabulous to taste. The recipes are healthful and taste great simply by adding herbs, spices, and extracts. Using herbs, spices, and extracts can enhance the flavor of food while taking out some of the fat.

Fat does add flavor and texture to food. Fat also helps to marry flavors and give a certain feel to foods. However, too much fat is also what makes us fat. So, we want to eliminate some of the fat from our diets. When we remove fat from a recipe, however, we need to replace it with something else that is going to add flavor or texture to make the food palatable. That's where herbs and spices come in. But, using too much of anything can be hazardous. For example, a mouthful of oregano or basil can leave you with the impression you've just eaten leaves and twigs. Herbs and spices need to be used subtly to bring out flavors without overpowering the base food.

When you go through this book, remember that it was designed for efficiency. Sometimes by spending time at the beginning of the month planning menus, you can save a lot of time in the long run. The same is true with shopping. Make lists for each week in the month and then buy just what you need when you need it. And make your kitchen a place where you want to spend time. A windowsill herb garden can add a wonderful, colorful, and fragrant touch.

For many of the recipes in this book, preparation time is under fifteen minutes. However, some of the cooking times are a bit longer. Read ahead in each recipe so you can plan your time. All of the recipes allow for some creativity on your part. Perhaps you prefer more or less pepper than the recipe calls for, or you wish to make a dish with more 'fire' by adding a touch of cayenne. That's your prerogative. I will give you some good guidelines and stimulate your ideas about how you can change your own favorite recipes to be healthful, fun to cook, and tasty.

I would like to hear from you about the things you like to make. Also, I would like to have some of your favorite family recipes, your mother's, your grandmother's, along with any questions you may have. You can write to me at Thomson Delmar Learning, 5 Maxwell drive, Clifton Park New York, 12065. Or you can email me at judygillard@hotmail.com.

I had always lived at sea level. I think I was destined to move to many different areas to give me more empathy for life in other parts of the United States or the world! The biggest change was moving to the mile-high city of Denver. I was terrified to try my baked items, but all my fear was in vain. I found in the baking isles of the grocery store flour that is for high altitude, which has quickly become a staple in my kitchen. I also use Bisquick Reduced Fat Baking Mix and have never once had an altitude problem with it. The one difference I did notice is that some things do take longer to cook at this altitude, for example, water boils at a lower temperature, so a four-minute egg takes five minutes to get the same results!

I have also included some recipes that are a bit more complex for special occasions; and, believe me, they are worth the extra effort.

About Herbs and Spices

The use of herbs and spices almost coincides with human history. It certainly predates written history. The Chinese were said to be among the first to discover the many uses of herbs and spices, both for medicinal purposes and as flavor enhancers in cooking. Chrysanthemums were originally grown for their medicinal properties and were a valued ingredient in a Taoist elixir. Today in China, chrysanthemums are still used in soups, salads, and teas, as well as in beautiful floral displays.

A first-century cookbook, attributed to Apicus, a Roman epicurean, features the use of herb combinations as flavor enhancers. One recipe for cooking artichokes includes fresh fennel, cilantro, mint, and rue, pounded together, then reinforced with pepper, lovage, honey, oil, and liquamen (a strong fish-based sauce that Romans used in place of salt).

Spices have been highly prized throughout history; by the ninth century, they were considered as valuable as gold or silver. Cloves and mace sold for about $18 a pound, and pepper was sold by the individual peppercorn.

Around 1699, an Englishman named John Evelyn wrote a book listing 73 salad herbs with details for using each herb. The book's title, *Acetaria: A Discourse of Sallets*, illustrates the traditional classification of herbs. "Sallets" were salad herbs, "pot herbs" were those cooked in large cooking pots, "sweet herbs" were flavorings, and "simples" were medicinal herb compounds.

Herbs are often coupled with spices, yet there is a difference. Herbs are the leaves of fresh or dried plants. Spices are the aromatic parts of the plant—buds, fruits, berries, roots, or bark, usually dried. An example of their relationship is coriander: The seeds of coriander are used in curries and chili powder. The leaves of the plant are known as cilantro, which is often used in Mexican cooking.

Herbs and spices define specific ethnic flavor preferences. In India, curry is created with as many as 10 herbs and spices. In Thailand, curry is used in conjunction with fresh herbs to give it a more delicate flavor. The Chinese use their famous five-spice powder along with ginger and garlic to give their food its distinctive quality. In Indonesia, flavor preferences tend to sweet and sour, and they use lemon grass, tamarind, Kaffir lime, and various chilies. In Europe, herbs are used sparingly with a focus on tarragon and the French *fines herbes*, a combination that includes parsley, chervil, and tarragon, among others. Greek and Italian cuisine emphasizes basil, thyme, sage, and oregano. In Mexico, cilantro is combined with various chili peppers and, more recently, epazote leaves to give a kick to refried beans.

In early America, almost every Colonial home featured an herb garden, but somewhere along the line we lost sight of the value of herbs in cooking. In 1939, Irma Goodrich Mazza wrote a best-selling cookbook, *Herbs for the Kitchen*, which reintroduced the use of fresh herbs to American cooking. Ms. Mazza reminded American cooks what fresh herbs, garlic, and premium olive oils could do to enhance the flavors of traditional American fare. She featured six herbs in her recipes: basil, marjoram, mint, rosemary, sage, and thyme.

During the 1970s and 1980s, Americans turned away from the kitchen and headed for fast foods. But now we have entered the 2000s, and we are back to cooking what we eat so we have more control over our health. We want to prepare healthful dishes, but we want to do it without much fuss. Interestingly, the herbs that are the most popular today are the six herbs Ms. Mazza featured, but we've also added parsley, cilantro, chives, and tarragon to the list.

Today, it is easy to have an herb garden. Fresh parsley can be grown in a pot on a windowsill, and mint does very well in the garden. As I describe these herbs and spices in detail, I will give some information on growing them. Check with your local nursery about the specific herbs that will grow well in your area. Remember, a kitchen garden today literally means a garden that will grow inside your kitchen. Have fun with herbs!

Harvesting, Drying, and Storing Your Homegrown Herbs

Some things you'll need to know about caring for herbs, in addition to the tips offered later on growing them, center on how to harvest them. Different parts of the herb are gathered at different times. Obviously if you are growing garlic, the entire plant will be taken at the time you're ready to use it. But annual leafy herbs, such as basil, should be carefully picked, never taking more than 10 percent of the growth at a time. The same is true with perennials like sage, thyme, and rosemary. Severe pruning or overstripping of the leaves will weaken the plant. Careful pruning or harvesting, on the other hand, results in more vigorous leaf growth, giving you healthier plants.

As a general rule, pick herbs when they contain the highest amount of flavor essence. Leaves should be picked just before the plant is about to flower. Flowers, on the other hand, are picked just before they reach full bloom. Berries and fruits are picked at their peak ripeness. When you are using the above-ground portion of a plant, pick just before the plant begins to flower. Roots, like garlic, or rhizomes, like ginger and turmeric, are collected in the fall, just as the leaves begin to change color.

When storing your herbs, keep in mind that herbal properties may be destroyed by heat, bright light, exposure to air, or the activity of plant enzymes, bacteria, or fungi. So, herbs should be kept in a cool, dry place, with minimum exposure to air and sunlight. This doesn't mean you should hide your collection of kitchen herbs and spices away where you forget to use them. It simply means that if you have a choice, put them in the cupboard instead of leaving them on the countertop.

One of the most popular methods of preserving herbs for use during winter months is drying. As a matter of fact, drying actually improves the flavor of some herbs, particularly bay leaves. Bay leaves should be cleaned using a pastry brush, but no water, and then laid out to dry in a warm place on an airy surface, like a screen. They dry in about a week and are ready for storage in airtight tinted-glass jars.

Other herbs may be dried in bundles. One easy method is to pinch together a small bunch and secure it with a rubber band or kitchen string. Hang the bunch upside down from a rack in a dry, somewhat cool location. The temperature of the drying area should not exceed 86°F because the essential oils of the herbs will evaporate at this temperature or higher. The kitchen is not really the best place to dry your herbs because of the added humidity from cooking. Try to find a spot that is relatively dry, or at least consistent in humidity.

There is yet another way to dry your herbs. Place fresh herbs in brown paper bags labeled for each herb type. Set the bags in a dry, dark, cool place until the herbs inside are dry and crunchy. Shake the bags occasionally so the herbs dry evenly. Remove any stems, and prepare herbs for storage by crushing the leaves or chopping them in your minichopper. Always store the dried herbs in airtight jars. Keep the jars away from light to protect the color and flavor of the herbs.

Remember, it doesn't take long for herbs to dry. Never let the leaves become so dry that they disintegrate into powder when they are touched. If there is no condensation in the jar by the next day, the herbs are ready to store.

Using the microwave oven to dry herbs is a quick and effective method. Remove the leaves from the stems after you have given the whole herb a quick rinse to remove any soil or dust. Be sure to pat the herbs dry before you strip the leaves. Then spread the leaves in a single layer between two paper towels, and microwave them on high for 2 to $2^1/_2$ minutes. Store the herbs in airtight tinted-glass jars.

Freezing is another effective means of storing herbs. Dill, fennel, basil, and parsley can all be frozen for future use. Clean the herbs and put about 2 or 3 tablespoons of each in separate freezer bags. You can freeze them alone, or you can make up bags of your favorite combinations. Be sure to label the bags so you can find the herb you want when you need it. Chopping the leaves and freezing them with a bit of water in ice-cube trays is another freezer-safe method of storing herbs. This is especially nice when you want to use the herbs in sauces and broths. Chop the herbs very fine and fill each cube half with the herbs and half with water, then freeze.

I put the frozen ice cubes into plastic bags and then place the bags in plastic freezer-safe boxes. That way the ice-cube trays are available for making ice cubes, and the extra packaging helps to retain the freshness of the herbs. I also like to store my bags of freezer-dried herbs in a plastic freezer-safe box to protect them from freezer burn or other damage and to make them easily accessible when I need them. Try to use frozen herbs within about six months.

The most important herb to have on hand always is fresh chopped parsley. Buy it in bunches, wash it, and dry it well. Put the parsley in your food processor with a steel blade, and run it until all the parsley is chopped. Then put the chopped parsley in an airtight container, and freeze it. You can easily take the amount you need as you need it.

Infused Oils

It is very helpful to keep a variety of infused oils for quick cooking. I like to keep infused flavored oils in a spray bottle to lightly spray items for added flavor and better end results in roasting. The ones I use in the recipes in this book are garlic and chile pepper oils. However there are many on the market, so experiment to find the ones you like most.

The Freshness Test

In most cases fresh herbs are really your best bet, but it's not always possible to get them or grow them, so when you use dried herbs, give them the freshness test by crushing them, using a mortar and pestle (a good mortar and pestle, by the way, is an essential kitchen item!). Crushing releases the flavor, enabling you to get the freshest taste out of the herbs when you add them to food. For the most part, you can expect herbs you have grown and dried yourself to last at least two years. Herbs you buy in the store may have been on the shelf for a while already, so test them for freshness when you use them. Herbs from your grocer will usually remain fresh only about a year in your cupboard.

Basic Herbs and Spices

Here's a basic list of herbs and spices that you might want to have on hand.

Allspice

Allspice is the dried, nearly ripe berry of the allspice tree, a member of the myrtle family. The name is derived from its flavor, which is pungent and sharply aromatic, and suggests a mixture of cloves, cinnamon, and nutmeg. Grown in the Caribbean, allspice is also known as pimento, pimenta, Jamaican pepper, or clove pepper. It is used whole or ground in pickling spices, mincemeat, roast meats, and baked goods. Its essential oil is used in meat sauces, ketchup, and spice blends for pickles and sausages, as well as for reproducing certain fruit flavors.

The allspice tree is a tropical tree with large, simple leaves and tiny flowers. Today, most of the allspice we use comes from Jamaica. Mexico also cultivates allspice for export, but the quality of the fruit is not as good as that grown in the West Indies. Attempts to grow the trees in the East Indies have failed, so allspice remains the one major spice produced exclusively in the Western Hemisphere.

Ground allspice from your local grocer can be kept up to two years. If you use the mortar-and-pestle test, you can assess whether your supply is still flavorful. You can also obtain whole allspice berries in 1-ounce containers at grocery stores or from herb dealers. For best results in cooking or baking, use $\frac{1}{4}$ to 1 teaspoon of ground allspice for every four servings, or use 3 to 6 whole allspice for the same yield.

Anise

Anise is an annual herb of the carrot family cultivated for aniseed, its small, fragrant fruits. It has a licorice-like flavor and is good in baking, in stews, and with vegetables. The extract is also good in espresso in place of liqueur. As a matter of fact, anise is the base of the popular Greek drink, ouzo. Anise is also known as aniseed, sweet cumin, star anise, and Chinese anise.

In the East, anise is used to flavor duck and pork dishes. In China, it is added to tea and coffee. Today, the Chinese star anise, a native of Southeast Asia, has replaced the more expensive aniseed. Although it can be grown in North America, anise prefers a high, sheltered, sunny location in a soil with good water retention. The plant is a shrub and can be incorporated into an herb garden in some locations. In the United States, its cultivation is limited to the Midwest and the East, especially Rhode Island.

In addition to the seeds, both dried and fresh leaves are used to flavor foods. The fresh leaves especially give a slightly sweet flavor to fruit and vegetable salads. The leaves are also a tasty addition to shellfish when placed in the boiling water in which the shellfish are cooked.

To achieve the best flavor, my rule of thumb is to use $\frac{1}{4}$ to $\frac{1}{2}$ teaspoon of the dried leaves, 1 to 2 teaspoons of chopped fresh leaves, or 4 to 6 whole leaves for a four-serving recipe. It's best to buy or harvest the seeds whole in small quantities because they do not retain their flavor long. Grind the seeds just before adding them to the recipe.

Basil

Basil, also called sweet basil, is an annual herb belonging to the mint family. Sweet and pungent, it is an excellent complement to tomatoes and cucumbers. Basil enhances the flavor of most cooked vegetables and is also the key ingredient in pesto, the popular sauce for pasta dishes. Basil is also an ingredient of *fines herbes,* an herbal mixture used in French cooking. It has a minty, mildly peppery taste and a rich aroma.

As a member of the mint family, basil makes a great potted plant in either the kitchen or the garden. It is also known as St. Josephwort, but there are actually 50 or 60 species of basil. The plant usually blooms in August, and the white blooms should be pinched off to promote leaf growth. The leaves are what is used in cooking. The varieties of basil that do well in America are sweet basil, dwarf basil, Italian or curly basil, lemon basil, and purple basil. Because it is a member of the mint family, basil should be contained in a pot, either in a window box inside your kitchen, or even in the garden. Left unchecked, the plant will spread into the growth plots of other herbs.

Although it is commonly available in the dried form from the grocery store, dried basil does not compare in taste to basil freshly picked from the garden or the potted plant. Basil is easy to grow and the fresh leaves can be kept briefly in plastic bags in the refrigerator or frozen with a little water in ice-cube trays.

In cooking, for each four servings use $\frac{1}{8}$ to $\frac{1}{4}$ teaspoon of the dried leaves, 2 to 3 teaspoons of the chopped fresh leaves, or one small sprig of fresh leaves from the plant.

Bay Leaf

Bay, or sweet laurel, a flavoring agent, is the leaf of the true laurel, a small evergreen tree or shrub native to the Mediterranean. It is cultivated in Greece, Portugal, Spain, and Central America. There is a form of the laurel tree, grown in California, which is a much larger tree whose leaves are used chiefly for their yield of volatile oil. Bay leaves were the laurels used for heroic Greek and Roman wreaths. The term *poet laureate* derives from the use of the bay or laurel leaf in the wreaths used to honor poets.

Whole or ground bay leaf is used to season meats, potatoes, stews, sauces, fish, pickles, and vinegars. However, its dry form is best in cooking. The fresh leaves have a slightly bitter flavor that dissipates if the leaves are left to dry for a few days. Even dried, bay has a strong flavor, and the leaves are sharp, so the leaf itself should be removed and discarded after it has flavored the food.

Bay is excellent in soups, stews, and marinades. It is frequently found in French cuisine and imparts a slightly sweet taste. However, it can leave a bitter taste if used too heavily, so I recommend the use of no more than 1 or 2 crushed leaves, or 1 to 3 whole leaves, in servings for four. Be sure to remove the leaves before serving the food.

Capers

Capers are the unopened flower buds of a deciduous shrub native to the Mediterranean. In Europe, the bush has been cultivated for its flower buds, which are picked, pickled, and sold as a pungent condiment. The shrub is grown as a greenhouse plant in the northern United States and outdoors in warmer areas.

Today, capers are found commercially packed in vinegar. Used frequently in Mediterranean cuisine, capers add a lot of zest to tomato dishes and eggplant and are especially good with fish. Capers are cured and prepared in salt, then put into a vinegar brine. Their bitter, salty taste makes them useful in small quantities.

The Italians place five or six capers on antipasto. Placing one caper on each canapé instead of sliced olives makes a distinctive change of flavors. For four servings, I use 1 to 2 teaspoons as a garnish or a single tablespoon in sauces.

Caraway Seed

Caraway is an annual or biennial herb, also from the carrot family. It is cultivated for its small, fragrant fruits, called caraway seeds. However, the fresh leaves and roots from your own herb garden can also be used, and each part of the plant has its own distinctive flavor. Caraway's feathery green leaves resemble the foliage of carrots and taste a bit like the seeds. The root is very sweet and somewhat like a parsnip, but much milder in taste. The small, elongated seeds are used in baking, desserts, cakes, and bread. The plants do well in temperate climates and require much the same soil and light conditions as the carrot. Roots and leaves store only briefly in the refrigerator. The seeds can last up to two years and should be stored in airtight containers and kept away from the light.

Caraway also goes well with meat, potatoes, and cabbage. It is often used in rye bread and with cabbage because it is believed to dispel gas and calm the digestive tract. Add caraway seeds to any dishes that might benefit from their unique flavor. Caraway is often used in soups, stews, cheeses, sauerkraut, and pickling brines. Caraway oil is also used to flavor two digestive-aid liqueurs, Scandinavian aquavit and German kümmel.

For four servings, use approximately 1 to 2 teaspoons of the chopped leaves and one to two fresh sprigs only as garnish. If you want to eat the root of the plant, plan one root for each serving. A teaspoonful of the seeds, crushed, enhances the flavor of baked apples, boiled potatoes, or cabbage. Start by using $\frac{1}{4}$ teaspoonful at a time to see what works best for your particular taste.

Cardamom

A member of the ginger family, cardamom is used in Indian and Middle Eastern–style cooking. Ground cardamom seed is used in curries and in pastries, buns, and pies. Its flavor is sweet, aromatic, and pungent. It is also used in espresso and to flavor coffees.

Cardamom pods are the dried fruit of a perennial native to India. Today, we import the seeds from Guatemala, Italy, Central America, Mexico, and Ceylon. It does not grow in North America. Next to saffron and vanilla, cardamom is the third most expensive spice available today.

Cardamom seed can be purchased in the pod or out of the pod, whole or ground. The ground or loose seeds lose flavor quickly; it is best to buy whole pods. Discard the pods themselves before grinding the seeds. Cardamom pods should be green or white, not brown. The brown pods yield a seed that tastes like moth balls, so beware!

The spice may be used much like cinnamon in baked apples, coffee cakes, melon balls, curries, pickles, honey, mulled wine, or grape jelly. Its flavor, which some describe as anise-like, also has a lemon tang. It enhances meats, game, and sausage. Use cardamom in small pinches to suit your personal taste. Try adding $\frac{1}{2}$ teaspoon ground cardamom to a fruit salad that serves four.

Cayenne

A member of the nightshade family, cayenne fruit, known as African pepper and African chilies, is picked when it has turned red. Then it is left to dry and is ground into a powder. Today, this spice is the ground pod and seeds of various chili peppers grown in Africa, Mexico, Japan, Nigeria, and the United States.

Cayenne flavors hot, spicy dishes, eggs, and beans. It takes very little cayenne to spice up a sauce or add zest to a chili. This seasoning resists measuring and must be used "to taste." Heat brings out the hotness of cayenne, so when you use it in cooking, add, stir, and taste. Then taste again in five minutes berore adding any more.

Celery Seed

A member of the parsley family, celery seed adds a wonderful flavor to foods to which you usually add celery. Its flavor is slightly bitter and it goes well with cheeses, spreads, cocktail juices, and pastry. Some cooks consider celery seed a prerequisite in potato salad. The seeds add the sweetly aromatic flavor of fresh celery, plus the slight, natural bitterness of the seed covering. Use in pastries or to flavor pot roasts, salad dressings, salads, sauces, soups, stews, and sandwich spreads.

Celery seeds are the dried fruits of the celery plant, a vegetable that is not well suited to home gardens because it requires very rich soil and up to a half year of cool temperatures to mature. Celery seed is available ground in $1\frac{1}{4}$-ounce containers at grocery stores and whole in $\frac{1}{2}$- to $1\frac{1}{2}$-ounce containers. Celery seeds are tiny and brown in color. The whole seed retains its flavor well, but it should be used sparingly because it can be bitter. If you want to maximize the flavor, crush the seeds before using them. Celery seasoning tends to become stale quickly, but you can refrigerate it to get better results. It should be kept in an airtight container in a cool, dark place.

For four servings, use $\frac{1}{8}$ to $\frac{1}{2}$ teaspoon of the ground seed or $\frac{1}{4}$ to 1 teaspoon of the whole seed as flavoring, or 1 teaspoon to $\frac{1}{2}$ cup of the whole seed in pickles.

Chervil

Akin to parsley with a slight likeness to tarragon, chervil is one of the four basic *fines herbes* found in the French kitchen (the others are parsley, chives, and tarragon). It is an annual herb of the carrot family, native to southeastern Europe and nowadays cultivated in Belgium and California.

Chervil is one of the best herbs for growing in boxes, and if you are lucky enough to have a greenhouse, you can count on having fresh leaves throughout the winter. It can be started from seed in your garden, then transplanted to pots for the winter. A sunny kitchen window can make a wonderful winter greenhouse for many of the herbs in this list, and you can enjoy them fresh throughout the entire year.

The leaves are best used when fresh, but you can refrigerate them in a plastic bag for a while. Although much of the flavor is lost, chervil can be dried and stored in an airtight jar.

Chervil can be used in soups, stews, salads, or in dishes in which you would use parsley. It imparts a mild anise-like flavor, like parsley but more subtle. Always add the leaves at the end of the cooking time to retain chervil's delicate flavor. The rule of thumb is to use $1/4$ to 1 teaspoon of the dried leaves for four servings. As a garnish, use 1 to 2 teaspoons of the chopped fresh leaves or six to eight sprigs for four servings.

Chinese Five-Spice Powder

This aromatic blend of spices gives a sweet, pungent flavor to roasted meats and poultry when they are cooked in soy-based sauces. Use it sparingly, however, since overuse can overpower the flavor of the main dish it is used to enhance.

Five-spice powder contains approximately equal portions of finely ground fennel seed, cloves, star anise, cinnamon, and Szechwan pepper. There is also a Chinese eight-spice powder that uses these five spices plus ground ginger, aniseed, and licorice. Because it is a blend, it is best to buy eight-spice from the grocery store or from an herb dealer.

Chives

The most delicate member of the onion family, chives enhance the flavor of many foods and may be used in stews, stock, or sauces, atop sour cream or cheese, added to omelets, or mixed in cottage cheese or cream cheese for spreads or dips.

Chives grow nicely in an herb garden and are often used as the decorative edging plant. They are a perennial and can be started from seed; but starting from the bulbs, or even better, buying the plants already started from the local nursery is your best bet. They do very well in kitchen windowsill gardens during the cold months and flourish outside from spring until fall.

Chives really cannot be dried with any great success, but they may be frozen and stored. However, because the plants are easy and undemanding, it is best to use their leaves fresh. To freeze, cut the stems and fill an ice-cube tray half with cut chives and half with water. To dry them, suspend chives individually by their flowered heads from a mesh screen. The fresh floral heads are edible and make a lovely garnish or component in salads.

As a rule of thumb, use 1 teaspoon to 2 tablespoons of the chopped fresh leaves for every four servings, or six to eight whole leaves in herb bouquets. If you want to use chives in herb butters, use 1 tablespoon to each quarter-pound of butter or butter substitute. For herb salad dressings, add 2 teaspoons to each cup of dressing. Because long cooking destroys their flavor, chives should be added at the last minute.

Cilantro

Cilantro is actually the leaf of the coriander plant, another member of the carrot family. It is one of the most widely used cooking herbs in the world. The leaves are shaped like parsley and have a pungent aroma and taste.

Cilantro grows well in gardens, preferring dry soil and full sun. Start the plant from seeds in the late spring and expect germination to be slow. The root of the plant can be cooked and eaten as a vegetable, and the seeds are used as a spice called coriander. By the way, the plant itself has a rather unpleasant odor, so you may want to buy this herb at the grocery store. If you do grow cilantro, dry it just as you would parsley; but bear in mind that the dried version is really not comparable to the fresh leaf.

Fresh cilantro does not keep well, so try to keep the fresh plant intact and store it in the refrigerator between moistened paper towels, or place the stem ends in a glass filled with water in the refrigerator. Remove any wilted leaves. Don't remove the roots or rinse the herb until you are ready to use it.

Fresh cilantro leaves may be frozen. Chop the leaves finely and place them in ice-cube trays—half cilantro leaves, half water. The seeds, available from the grocery store and known as coriander, should be kept in a cool place away from light in an airtight container.

Cilantro is good with poultry, vegetables, and sauces. It is commonly used in Chinese and Mexican cooking and is also known as Chinese parsley. Use it almost the same as parsley, but a little more sparingly: one or two sprigs on each plate as garnish or about $3/4$ tablespoon of fresh-cut leaves for every four servings.

Cinnamon

Cinnamon is one of the oldest spices known to humans. It is used either as the whole bark or in ground form, mainly in baking and pickling. Sprinkle it on toast, add it to cookies, stir it into hot apple cider or cold applesauce. Cinnamon is also known scientifically for its antiseptic properties, which explains why it is found in mouthwash and toothpaste.

This spice is actually the dried inner bark of trees belonging to several species, all related to the laurel family. The true cinnamon tree is native to India and Sri Lanka and is cultivated in various locations in the tropics.

Cinnamon is available from the grocery store ground in $1\frac{1}{2}$-ounce to 4-ounce containers, or in quills or sticks in about $1\frac{1}{2}$-ounce containers. The spice usually used in the United States is really cassia, the dried ground bark of the trees from the laurel family. It has a stronger flavor than true cinnamon and is a deeper brown; real cinnamon is actually a light yellowish brown. Store cinnamon sticks or ground cinnamon in airtight containers in a cool, dark location.

The rule of thumb for using cinnamon is to add $\frac{1}{4}$ to 1 full teaspoon of ground cinnamon per four servings or 1 small quill or stick per serving. Try cinnamon sprinkled over cooked squash or on peas or spinach. Adding a pinch in chili powder really brings out the flavor. For an unexpected taste, add a pinch of ground cinnamon to a meat stew.

Cloves

Cloves are actually the dried buds of an evergreen tree of the myrtle family. The tree produces abundant clusters of small red flower buds that are gathered before opening and are then dried to produce the dark-brown, nail-shaped spice we recognize as cloves.

When you step into a spice shop, it is the rich, warm aroma of cloves that usually greets you. Cloves have been used throughout history to sweeten the breath. Buy cloves ground in $1\frac{1}{4}$-ounce containers or whole in 1- or 2-ounce containers. Store them in airtight containers in a cool, dark place. Expect them to last on the shelf about a year.

Cloves may be used whole, wrapped in cheesecloth for easy removal after cooking, or ground in drinks, marinades, glazes, breads, and sweets. Ground cloves are often used in combination with a bay leaf, or ground cinnamon, ginger, and nutmeg. Cloves have a very strong flavor, however, so it's best to be conservative until you are really comfortable with using the ground form.

Whole cloves are often used to stud a ham. They may also be added to the water in which vegetables are boiled or steamed to give a wonderfully warm flavor. Using a whole onion studded with cloves during the last hour of cooking a roast imparts a delicious taste.

For four servings, use $1/8$ to $1/2$ teaspoon of ground cloves. In beverages, 1 to 2 whole cloves are best. With roasts or hams, the amount varies for either appearance or taste.

Cumin

Cumin seed is actually the dried fruit of a small annual herb that belongs to the parsley family. This delicate plant grows in Egypt, western Asia, and the Mediterranean. Often found in Mexican and Indian cooking, cumin was used in Roman times the way we use pepper today.

Ground into a powder, cumin is hot and bitter, and should be used sparingly. For four servings, use $1/8$ to $1/2$ teaspoon of ground cumin or $1/2$ to 2 teaspoons of the whole seed. The seed is yellowish brown and oval shaped, something like a caraway seed. You will find cumin more flavorful if you use whole seeds and grind them with a mortar and pestle just before use. Their flavorful oil dissipates rapidly after they are ground. Dry roasting the seeds before grinding enhances their pungent flavor. Remember, cumin is potent and can dominate the taste if used too generously.

Cumin is used commercially in flavoring liqueurs and cordials. Use it in curries and for pickling. Although ground cumin seed is already a part of chili powder, many Mexican cooks like to add a teaspoon of whole cumin seed per pound of meat in their chili con carne.

Cumin can be grown in your herb garden, but it is fragile. It is an annual that can be started from seed sown in the late spring. It likes sandy soil and warm weather and tolerates most well drained soils in sunny settings. Because only the seed is usable, it is probably more efficient to buy cumin. It is available at the grocery store in $1/2$- to 1-ounce containers, ground, or whole in $1 1/2$-ounce containers.

Curry Powder

Curry powder is actually a blend of ingredients including turmeric, cardamom, coriander, mustard, saffron, allspice, and other spices. It can be used in salad dressings, in egg dishes, on poultry, and to make a distinctive savory rice. It is most often associated with the flavors of India.

In the kitchen, you may want to proceed slowly with the use of curry until you are familiar with your blend. You may mix it yourself according to the following recipe, but for your first use, it's really best if it is purchased. Curry may be mild, somewhat hot, or very hot. A small touch of the powder in cranberry sauce to go with roast chicken or a bit in baked fish might be a good way to start.

To make a basic curry blend, place 6 dried red chilies, 2 tablespoons coriander seeds, $1/2$ teaspoon mustard seeds, 1 teaspoon black peppercorns, and 1 teaspoon fenugreek seeds in a heavy skillet. Roast them over a medium heat until they are dark in color. Be very careful not to burn the mixture. Leave this to cool, then grind it into a powder with a mortar and pestle. Blend in $1/2$ teaspoon ground ginger and $1/2$ teaspoon ground turmeric. Curry can be stored up to three months in an airtight jar kept in a cool, dark place.

Dill

Dill has long been a favorite flavor in pickles. It can be found in herb form (dillweed) or as dill seed, used as a spice. A member of the carrot family, dill is an easy plant to grow and makes another good addition to an outdoor herb garden. Dill is an annual that tolerates most soils. Sow the seed in the spring, then simply wait for harvest.

Dill is one of the herbs you want to use fresh if at all possible; however, dried dillweed works very well. I much prefer the dillweed to dill seed.

Fresh dill can be kept in a plastic bag in the refrigerator for a week or so, or you can freeze it. Add dill to whipped nonfat cottage cheese to create a delicious dip for vegetables. It is also very good sprinkled on seafood or meat to achieve a butterlike flavor. Add dillweed in tuna salad or in a potato salad along with fresh sliced cucumbers for a refreshing change.

Add 2 teaspoons of chopped dill mixed with fresh vegetables in salads. In cooking chicken, lay a single dill stem over the poultry before cooking. Two teaspoons of chopped dill with eggs makes a tasty omelet. Fresh dill makes an appealing garnish.

Fennel

Fennel is a member of the plant family that includes carrots and parsley. It has a sweet, licorice flavor similar to anise but is a bit milder. Throughout history, fennel was prized for its stems; today, we use it almost exclusively as a seed.

Fennel can be grown in the garden and is a perennial, but it can be cultivated as an annual. It may be planted in any soil except clay and the seeds should be sown in the autumn. Fennel has deep green fernlike leaves something like dill, and the large stems, flattened at the base, look a little like celery. Both the foliage and the stems have a mild licorice flavor.

A good addition to pork, fish, or seafood dishes, fennel seed may also be used in stocks, soups, and chowders, as well as in rice dishes. Use 1 teaspoon to 2 tablespoons chopped fresh leaves as a garnish for four servings, or use $1/2$ to 1 teaspoon of the seed, crushed.

If you grow fennel in your garden or buy it fresh at the grocer, you can eat it raw, just as you would celery. Fennel is a pleasant addition to a salad.

Garlic

Used since ancient times, garlic is a perennial herb related to the onion. It is a pungent bulb composed of cloves surrounded by a thin white or purplish sheath. Because seed is rarely produced, garlic is propagated by planting the individual cloves. When the green tops ripen and fall over, pull the bulbs. Garlic can be stored for several months if it is kept dry and cool.

Most of the garlic in the United States is grown in California, and much of the crop is dehydrated and sold as garlic powder. If you try it in your garden, you'll want to know that it is a perennial that prefers rich, light, well drained soil. Plant the individual cloves in the spring or autumn in rich, dry soil, in a sunny spot. Put the bulbs down about 4 inches.

When harvesting garlic, you will be taking the entire plant, so be ready to replant with another bulb or clove. Fresh garlic is at its best at the beginning of the season and generally should be used as it is taken from the garden. If you want to store it, put the garlic in a cool, dry, well-ventilated place away from light. When stored properly, the bulbs should last several months.

Garlic adds a strong, pungent flavor to any dish. It is recognized for its medicinal and curative properties and is used largely in Italian and Mediterranean cooking. It is available in cloves, powdered, and as a salt. Garlic can be added to almost any food. The cloves' papery skins peel easily if you smash them with the flat side of a knife.

An alternative method for cleaning and using garlic is to plunge the cloves into boiling water for 30 seconds. Drain and then peel them when they are cool. Crush the cloves with the flat edge of a knife, then slice or chop them as required. Of course, using a garlic press is an excellent method for getting all the flavor from a clove. To be sure each batch of garlic contains only the freshly pressed pieces, however, you must thoroughly clean and dry the press after each use. Use a toothbrush to scour the press.

Garlic is available minced and chopped in jars in the grocery market, which makes it very easy to use. Olive oil–infused oils are also a welcome addition in many of my recipes. However, when time allows, nothing beats the flavor of fresh garlic. For four servings, use $1/2$ to 1 full garlic clove in cooking. When you want just a hint of garlic, rub a crushed clove around the base of a porcelain baking dish or a wooden salad bowl before you add anything else.

The latest fad in cooking is roasting garlic. I do believe this fad is here to stay. Roasting causes garlic to take on a whole new flavor, sweet and nutlike. Roasted garlic has a multitude of uses, from spreading it on toasted bread to adding it to a sauce or rubbing it on chicken before roasting. To roast garlic, choose very fresh rock-hard heads, then cut off the top and rub the head with olive oil. Wrap it in aluminum foil and roast in a 350°F over for 45 minutes. If you have a terra-cotta garlic roaster, soak the cover for 15 minutes in warm water, then place the garlic in the roaster. Cover the garlic and place it in an unheated over at 350°F for 45 minutes. Uncover the roaster and continue roasting for another 15 minutes. This produces a soft caramelized result. The cloves will pop out of their skins. You can use them right away or put them in a sealed container and keep it in the refrigerator for up to four days. You will find that roasted garlic will become a mainstay in your home.

Ginger

Ginger is a spice from a perennial plant native to southern Asia. It is found in powdered form or in the produce department as a whole root. Ginger adds a sharp edge to food. It is used in baking and in Oriental cuisine.

Ginger ale and gingerbread both evolved from ginger's use in ancient Greece as a digestive aid. The Greeks ate ginger wrapped in bread after big meals. Today, ginger ale is a popular drink known for its stomach-soothing properties. Throughout history, ginger has been used for its medicinal properties. In China, it is still considered an important drug.

Ginger is available ground in $1/2$-ounce, 1-ounce, and 2-ounce containers at the grocery store. The root—crystallized, dried, and sugared—comes in 8-ounce and 1-pound tins at confectioneries or herb dealers. The root also comes preserved in syrup in jars of varying shapes and sizes. The whole dried root may be purchased in 1- or 2-ounce containers at grocery stores and markets.

If you buy the fresh root, wrap it in paper towels and seal it in a plastic bag, then refrigerate. It should keep for several weeks. The dried product should be kept in a cool, dark place. Use a sharp knife to prepare your fresh gingerroot for grating. Peel away the tough outer skin only as far as the flesh, then use a ginger grater.

For a different twist on a filet mignon, try blending $1/2$ teaspoon of ginger with salt and pepper, then rub the seasoning mix on both sides of the meat before broiling. Poultry and pot roasts are also delicious when lightly sprinkled with ginger before cooking. Remember, dried ground ginger tastes nothing like fresh ginger and the two are not interchangeable in recipes.

Try including ginger in pies, pickles, puddings, sauces, and baked fruits. The rule of thumb for use in cooking is to add $1/4$ to 1 teaspoon of ground ginger for each four servings. The amount of fresh ginger you use will vary from recipe to recipe. Add a little at a time to determine the amount that suits you best.

Leeks

Leeks are an annual herb belonging to the lily family. The leek's onionlike flavor is mild and sweet. Leeks look like a very large green onion. They can be somewhat sandy, so they need to be washed well. They can be grown in your garden in a mound of earth surrounded by a trench. The herb develops long stalks underground and broad, flat green leaves above. Leeks are always used fresh.

Leeks add a delicate, light flavor to food and are frequently found in French cooking. They enhance the flavors of meat, fish, and poultry dishes and may be used as a garnish. Added to pasta, rice dishes, and salads, leeks impart a slightly oniony taste. Use $1/2$ to a whole leek in salads and soups for four. You can serve leeks by themselves as a cooked vegetable.

Leeks make an ideal base for a fresh bouquet garni. *Bouquet garni* is the French term used to describe a bundle of herbs to be used in cooking. The classic combination consists of 3 sprigs of parsley, 1 small sprig of thyme, and 1 small bay leaf, wrapped in an aromatic vegetable wrapping, like the green part of the leek or a celery stalk. The elements of the bouquet are tied together with string, then added to the soup, stew, or sauce. When the food has cooked, the garni is removed.

There is a dried version of this blend that can either be purchased or mixed at home. Use equal parts of dried herbs, such as parsley, bay, and thyme, or whatever combination you prefer. Place the herbs in a square of cheesecloth and tie the four corners with kitchen string to form a bag. Remove the bag after cooking. Choose herbs suitable for the dish you're making.

Lemon Zest

This is the thin outer peel of the lemon. Although fresh is best, lemon zest is available in jars and is good to keep on hand. I've included lemon zest in many of the recipes in this book. It imparts a slightly bitter, citrus flavor and is usually used in small quantities.

Mace

The ground outer coating of the nutmeg seed, mace is used in the same ways as nutmeg. Nutmeg fruits resemble small peaches and are quite beautiful inside when ripe. The inner layer is green, the next layer is orange, and the outer layer, which becomes the covering that is dried and ground to produce mace, is a brilliant scarlet. When dried, it forms a lacy outer covering over the nutmeg kernel. More than 400 pounds of this covering are needed to make one pound of mace.

Ground mace stores better than almost any other ground spice. Keep it in an airtight jar in a cool, dark place and it will last for years.

Commonly found in cakes, cookies, pickling, and preserves, mace can also be used in some meat and fish dishes. It blends well with cinnamon, cloves, allspice, and ginger. Mace's flavor is much stronger than that of nutmeg, and in blends, you'll want to use less than any other component. When making a cream sauce or a fish sauce, for example, use $1/8$ to $1/4$ teaspoon of ground mace and $1/4$ teaspoon of onion salt with $1/4$ teaspoon of celery salt. For shellfish, add a few grains to each serving. The rule of thumb is to use $1/8$ to $1/2$ teaspoon of ground mace for four servings.

Marjoram

Marjoram, or sweet marjoram, is a perennial that grows to be about 12 inches tall and has grayish green leaves and tiny cream-colored flowers. It is native to western Asia and the Mediterranean region, but you can try to grow it in your herb garden. Sow the seed in late spring or early summer (or start the plants indoors and replant them outdoors if you live in a colder climate) in medium-rich, finely prepared soil. The plants prefer sunny locations and damp-to-dry soil. Keep marjoram in the refrigerator in sealed plastic bags when it is fresh.

Similar to oregano, but milder and sweeter, marjoram is very good in sauces, stews, and soups. It is also good over vegetables. Many cookbooks suggest replacing oregano with marjoram for sweeter, spicier sauces. But the fact is, the oregano on your spice rack may be marjoram. All marjoram species are also called oregano. Although some palates cannot distinguish between the two, there is a difference. When you want milder oregano flavor, use marjoram.

The rule of thumb for four servings is $1/4$ to $1/2$ teaspoon of crushed dried leaves, or three or four small fresh leaves.

Mint

Several different flavors of mint can be grown easily in the garden. In fact, they grow so well they just might take over. When planting mint in the garden, plant it in pots and then place the pots in the soil to contain the growth. You can also plant mint in a bricked-in section of your herb garden as was often done in Colonial times. Mint is a perennial but should be replanted about every four years because it can become woody.

Mint is actually the common name of about twenty-five perennial species of one family. The common garden mint is spearmint, which has a sweet, strong scent and is widely used in candies, chewing gum, and herbal teas. Peppermint is often found growing wild in damp locations from Nova Scotia west to Minnesota and south to Florida and Tennessee. Other common varieties include lemon mint, water mint, and round-leafed mint.

Mint is really best when used fresh from the garden. If you want to store it, however, plan to store it only briefly. If you have planted mint in pots in the garden, you can bring the plants into the kitchen in cooler weather and enjoy fresh mint all year long. If you need to dry some, use the hang-dry method and keep the dried leaves in airtight containers in a cool, dry location. You'll lose a lot of flavor with the drying process, so adjust your recipes accordingly.

Mint is one of the most widely used herbs in the world. Chop the leaves and add them to salads and cold drinks. Mint leaves also make a soothing hot tea. Used in cooking, mint enhances the flavors of meat and fish dishes; mint jelly is often served with roasted lamb. Minced mint leaves mixed with cream cheese make a tasty appetizer. Try using 2 tablespoons of freshly chopped mint over baked or broiled fish as a garnish and for flavor. Add 1 or 2 tablespoons of fresh mint to boiled carrots, green beans, peas, potatoes, or spinach for a delicious flavor variation.

Mustard

Mustard is the common name of some annual plants that are grown for their pungent seed and for their leaves, known simply as mustard greens. Mustard is native to all of Europe and southwestern Asia and is now cultivated in Austria, England, Germany, Holland, and the western United States.

Ground mustard or mustard seed is a wonderful addition to salad dressings or egg and vegetable dishes. The ancient Greeks and Romans ground up mustard seeds and spread the powder over their food as a spice. The condiment as we know it today was first produced in the seventeenth century. Today Dijon, France, produces over half the world's supply of prepared mustard.

White mustard seeds can be sprouted indoors or in the garden for a crisp and peppery salad herb. Place the seeds on a thin layer of soil in small trays, or even on a piece of damp cloth. Keep them moist and watch them sprout. It takes only about two weeks to get sprouts that are ready to eat (they should be about 2 inches high). Use these delicate sprouts in salads, on sandwiches, or as a plate garnish.

Store mustard seeds in a cool, dry place, preferably in airtight containers. If you grind the mustard seeds and then add a little water, you'll get a paste that is much like the condiment you buy as prepared mustard, but with a fresher taste.

Use mustard, ground or in seed form, to obtain a more subtle flavor than the condiment. Use $\frac{1}{8}$ to 1 teaspoon of seed for four servings, according to the taste you prefer.

Nutmeg

Nutmeg is the seed of the nutmeg tree native to the Spice Islands of Indonesia. The dried coating over the center kernel of the nutmeg, which is the pit of a yellow, apricot-like fruit, is known as mace.

Nutmeg has a wild, nutty, distinctive flavor and is very fragrant. It is very good in desserts and for baking and can be purchased as a ground powder. You can also buy it whole and grind it with a special nutmeg grinder or grater. Early spice merchants often carried nutmeg in a special container with a grater attached so the spice could be ground fresh when needed.

Nutmeg goes well with rich foods. It is used in filled pastas in Italian cuisine, either mixed into the stuffing or ground on the top at the last minute. Nutmeg is known to particularly complement milk and cheese dishes. Cabbage, kale, cauliflower, spinach, and sweet potatoes all benefit from the flavor-enhancing quality of nutmeg. Fruit salad with a tablespoon of vanilla yogurt and a few grinds of fresh nutmeg makes a lovely difference.

Nutmeg is also known medicinally as an aid to digestion, so it often appears in desserts served at the end of a meal.

Orange Zest

The outer skin of the orange, orange zest is best when used fresh. Keeping a jar of dried zest on hand, however, is helpful for those times when you want to be creative in the kitchen, but you don't have the fresh fruit available. Orange zest is a little less bitter than lemon zest. It's availale from your local grocery store in $1^1/_2$-ounce jars and can be kept on the shelf for about a year.

Oregano

Oregano is really a form of wild marjoram and a member of the mint family. It can be started from seed and grown in an herb garden as a perennial. Sow oregano in late spring on a warm site where the soil is nutrient rich and dry. Oregano has beautiful miniature lavender flowers and grows into a leafy bush about $2^1/_2$ to 3 feet tall.

Oregano has a somewhat stronger flavor than marjoram, but its flavor varies depending on soil and climate. Use oregano when you want a zesty taste. Use marjoram when you want a milder, subtler flavor. Italian and Mexican cooking make full use of oregano. As a matter of fact, the word *oregano* is Spanish for marjoram. Oregano is also known as "Mexican sage" because of its wide use in Mexican cuisine.

The dominant herb in most Italian pasta sauces, oregano is excellent when used with tomatoes, peppers, zucchini, eggplant, and in dips. Add about $1/_2$ teaspoon of dried oregano to spice up a potato salad, and $1/_2$ teaspoon of the dried herb is about right for plain tomato sauce. Use a full teaspoon of minced fresh oregano when it's available from the garden.

Paprika

Paprika is a mild, reddish pepper that is good with chicken, eggs, and dressings. It also makes a nice garnish to add color to any light-colored dish. Poultry, shellfish, salads, rice, fish, scrambled eggs, macaroni, mashed potatoes, appetizers, soups, and gravies all may be flavored and garnished with paprika.

Buy the best quality of paprika you can find, and be careful, some of it can be extremely spicy. Spanish paprika, for example, is really more like cayenne. Keep paprika in an airtight container in a dark, cool location in your kitchen. It loses its flavor and aroma quickly and becomes brown in color and stale to the taste if kept too long.

For four servings, use 2 teaspoons for mild flavoring and a tablespoon or two for a more pungent taste. As a garnish, vary amounts according to the appearance of the dish. By the way, paprika was found to contain more vitamin C than any citrus fruit, so you'll be glad to know it adds more than just color to your food.

Parsley

Parsley is a biennial herb of the carrot family that comes in more than 30 varieties. It is easy to grow indoors in a pot or outdoors in an herb garden. Plant the seeds anytime from early spring to early summer. Germination is slow but can be speeded up by watering the freshly planted seeds, which are drilled down.

Herb and Spice Chart

Here is a guide for seasonings for you to use when you are standing in front of your herbs and spices trying to figure out which one to use on what! This information came from McCormick & Co., Inc., one of the foremost spice companies in the United States.

White Vegetables

Cauliflower	Seasoned salt, salt 'n spice, white pepper, paprika, parsley, chives, rosemary
Potatoes	Dill weed, paprika, parsley, seasoned salt, white pepper

Red Vegetables

Beets	Garlic powder or salt, cloves, allspice, bay leaf
Eggplant	Marjoram, oregano, basil, black pepper, garlic powder or salt, italian seasoning
Tomatoes	Basil, oregano, marjoram, black pepper, dill weed, thyme, garlic powder or salt, Italian seasoning

Green Vegetables

Asparagus	Lemon & pepper seasoning salt, nutmeg, dry mustard, black paper
Broccoli	Lemon & pepper seasoning salt, black pepper, ground red pepper, garlic powder or salt, oregano
Brussels Sprouts	Lemon & pepper seasoning salt, black pepper, dry mustard
Cabbage	Garlic powder or salt, caraway seed, black pepper, thyme

Cucumbers	Dill weed, basil, black pepper, chives, tarragon, seasoned salt
Green Beans	Garlic powder or salt, black pepper, thyme, tarragon, dill weed
Peas	Minced onion, mint flakes, basil, chives, black pepper
Spinach	Lemon & pepper seasoning salt, nutmeg, mace
Zucchini	Oregano, black pepper, basil, Italian seasoning, marjoram

Yellow Vegetables

Carrots	Ginger, dry mustard, dill weed, chives, Season seasoned salt, lemon & pepper seasoning salt, black pepper
Corn	Chili powder, onion powder or salt, chives, black pepper
Sweet Potatoes or Yams	Cinnamon, nutmeg, allspice, mace, cardamom
Yellow Squash	Onion powder or salt, basil, dill weed, parsley, paprika, chives, black pepper
Winter Squash	Cinnamon, nutmeg, pumpkin pie spice, thyme, marjoram

Poultry

Chicken and Cornish Hens	Basil, dill weed, ginger, nutmeg, oregano, marjoram, thyme, celery seed or salt, chives, bay leaf, garlic powder or salt, onion powder or salt, dry mustard, parsley, paprika, rosemary, sage, tarragon, poultry seasoning, seasoned salt, Italian seasoning, saffron, sesame seed
Duck	Onion powder or salt, thyme, black pepper, rosemary, bay leaf, ginger, dry mustard
Turkey	Poultry seasoning, Season-A110 seasoned salt, black pepper, paprika, garlic powder or salt, onion powder or salt, sage, oregano

Gamebirds	Thyme, lemon & pepper seasoning salt, dry mustard, celery salt, parsley, paprika

Fish

Very Delicate Flavored fish	Parsley, lemon & pepper seasoning salt, peppercorns, bay leaf, tarragon, dill weed, white or black pepper, chives, basil
Light to Moderate Flavored Fish	Tarragon, garlic powder or salt, white or black pepper, oregano, dill weed, bay leaf, fennel seed, ground red pepper, Italian seasoning
More Pronounced Flavored Fish	Basil, thyme, marjoram, garlic powder or salt, black pepper, tarragon, parsley, dill weed, bay leaf, fennel seed, ground red pepper, oregano, rosemary

Shellfish

Clams	Black pepper, seafood seasoning, minced onion, parsley, garlic powder or salt
Crab	Crab and shrimp boil, dry mustard, seafood seasoning, dill weed, ground red pepper, tarragon, black pepper, parsley
Lobster	Black pepper, dry mustard, lemon & pepper seasoning salt, chives, tarragon
Mussels	Bay leaf, chopped onion, parsley, thyme, black pepper, garlic powder or salt, ground red pepper, oregano
Oysters	Paprika, parsley, Season-All® seasoned salt, thyme, fennel seed
Scallops	Dill weed, tarragon, lemon & pepper seasoning salt, white pepper, chives, dry mustard, paprika
Shrimp	Crab and shrimp boil, bay leaf, thyme, garlic powder or salt, oregano
Crayfish	Lemon & pepper seasoning salt, fennel seed, black pepper

Meat

Beef	Black pepper, thyme, oregano, marjoram, bay leaf, garlic powder or salt, onion powder or salt, chili powder, Season salt, meat tenderizer, parsley, cumin, ground red pepper, rosemary, dry mustard, ginger, curry
Game	Black pepper, marjoram, thyme, bay leaf, juniper berries, garlic powder or salt, onion powder or salt, meat tenderizer, Season salt
Lamb	Rosemary, garlic powder or salt, thyme, marjoram, oregano, Season salt, bay leaf, curry, celery seed or salt
Pork	Black pepper, basil, dill weed, thyme, oregano, marjoram, sesame seed, garlic powder or salt, onion powder or salt, Season salt
Veal	Parsley, dill weed, basil, lemon & pepper seasoning salt, paprika, tarragon, chives, celery seed or salt

Extracts

Liquid extracts are one of the newest forms of preserved herbs and spices. The essential oils of the herbs and spices are combined with alcohol and water to form a concentrated seasoning. Extracts may be added to foods just before serving. Since extracts are highly concentrated, add them drop by drop until you achieve the exact flavor you want. Here is a list of the most common extracts to keep on hand:

- Almond
- Brandy
- Orange
- Vanilla
- Anise
- Lemon
- Rum

Extracts can enhance the flavor of even the most basic dish. Try adding different extracts to coffee, for example, to make flavored coffees without adding alcohol. Extracts can be stored at room temperature and will keep for an extended period if they are carefully and tightly covered after each use. Naturally, some evaporation will occur over time because of the alcohol base used.

Remember, extracts can be overpowering because they are highly concentrated, so use them sparingly at first. Try using 5 to 10 drops in servings for four, but personal taste will govern your use. Whenever possible, use pure extracts rather than imitation, especially vanilla.

Herb Blends

Herb blends can be customized to your individual taste. The blends listed here are all made with dried herbs and should be mixed in a minichopper. Herb blends make nice gifts if you put them in small containers. A small, interesting airtight jar of your personal herb blend coming directly from the warmth of your kitchen is an environmentally friendly and usable gift, sure to please.

For these herb mixes, I used "parts" so you can create whatever size batch you need. For your own kitchen, I suggest you use 1 tablespoon as the part. If you are blending herbs to use for gifts, you might want to use a cup or a half cup as the part, depending on how much of the blend you want to give away. Some recipes in this book include these blends.

Basic Herb Blend
4 parts parsley
2 parts chopped chives
2 parts dillweed
2 parts oregano
1 part rosemary
1 part thyme

Chicken Blend
2 parts marjoram
1 part basil
1 part parsley
1 part dillweed
1 part paprika

Dip Blend
4 parts dillweed
1 part garlic powder
1 part chervil

Greek Blend
2 parts garlic powder
1 part lemon peel
1 part oregano
½ part ground black pepper

Herb Salt
2 parts onion powder
1 part garlic powder
1 part dry parsley
1 part marjoram
1 part salt
½ part basil

Italian Blend
2 parts basil
2 parts marjoram
1 part garlic powder
1 part oregano
½ part thyme
½ part rosemary
½ part crushed red pepper

Mexican Blend
1½ parts cumin
1 part onion powder
1 part garlic powder
½ part ground ginger
½ part paprika
½ part oregano
½ part dry mustard
¼ part cayenne pepper
½ part parsley flakes

Vegetable Blend
1 part marjoram
1 part basil
1 part chervil
½ part tarragon
½ part celery seed

Starters

A Word on Starters

SPREADS

Cheddar Cheese
Cheddar-Olive
Feta Cheese
Goat Cheese
Vegetable Cream Cheese

DIPS

Eggplant Caviar
Guacamole
Hummus

PÂTÉS

Chicken-Liver
Country
Spinach and Chicken

OTHER

Artichokes with Shrimp
Chicken Drumettes
Crab Cakes
Phyllo Goat-Cheese Squares

A Word on Starters

Starters can become the whole meal for an informal gathering of two or one hundred and two. It is always smart to have something to nibble on if you are serving cocktails and to be prepared for spur-of-the-moment entertaining. I always try to have a cheese spread made up in the refrigerator. Cheese spreads are easy to make and taste great. You can also reduce the fat content by mixing the real cheese with hoop cheese or dry-curd cottage cheese. Mix the two well, and the end result is all pure, with a good consistency and no plastic taste! Depending on how fast you use the spreads, it is nice to have two varieties at a time. They will keep as long as the "sell-by" date on the hoop or dry-curd cheese. Cheese spreads make a good snack and are at the ready for other needs, such as stuffing celery or spreading on crackers. The eggplant caviar is another item you could freeze in sealed containers to have on hand for spreading on crackers.

When entertaining a group for a walk-around meal, try the pâtés along with the Chicken Drumettes and the Crab Cakes. The only item that could be difficult is the Stuffed Artichokes, which are best served at a sit-down dinner as a first course. Add some interesting breads and cut-up vegetables, sliced apples and pears, and bunches of green and red seedless grapes. This could also be a perfect opportunity to do a wine pairing with the different foods. Add a table of desserts, coffee, and a few different Ports, and you have the end to a lovely affair.

Cheddar Cheese Spread

Serves 6

The most basic of all the spreads, Cheddar Cheese Spread is good on crackers and celery, and great with apples. Try coring an apple and filling it with the spread; then refrigerate the apple to make sure the spread is nice and firm. To serve, slice the apple so you have a round apple slice with a cheese center. The cayenne brings the flavor up a bit.

8 ounces hoop cheese or dry-curd cottage cheese
4 ounces cheddar cheese, grated
Dash cayenne pepper

Put all ingredients into a food processor with a steel blade and process until smooth.

Nutritional Analysis (per serving)

96	Calories
6.3 gm.	Fat
21.2 mg.	Cholesterol
58%	of Calories from Fat
8.8 gm.	Protein
4.01 gm.	Saturated Fat
0.0 gm.	Fiber
0.7 gm.	Carbohydrate
118.9 mg.	Sodium
142 mg.	Calcium

Cheddar-Olive Spread

Serves 6

Cheddar-Olive Spread has always been a favorite of mine. It is good to spread on crackers, fill celery, or make an old-fashioned grilled cheese sandwich with a new twist. If you have them, add extra pimentos.

8 ounces dry-curd cottage cheese or hoop cheese
4 ounces cheddar cheese, shredded
4 Spanish olives (stuffed with pimentos)
1/8 teaspoon crushed red pepper

1. Put cheese into a food processor and process with a steel blade until smooth.
2. Drop olives into the processor while it is running. Process until the olives are just chopped.
3. Transfer to an airtight container and refrigerate.

Nutritional Analysis (per serving)

99	Calories
6.6 gm.	Fat
21.2 mg.	Cholesterol
60%	of Calories from Fat
8.9 gm.	Protein
4.05 gm.	Saturated Fat
0.0 gm.	Fiber
0.7 gm.	Carbohydrate
182.9 mg.	Sodium
144 mg.	Calcium

Feta Cheese Spread

Serves 6

For variety, add diced sun-dried tomatoes and basil to the feta spread.

8 ounces hoop cheese or dry-curd cottage cheese

4 ounces feta cheese

Dash cayenne pepper

Place all ingredients into a food processor with a steel blade and process until smooth.

Nutritional Analysis (per serving)

70	Calories
4.1 gm.	Fat
18.2 mg.	Cholesterol
52%	of Calories from Fat
6.8 gm.	Protein
2.85 gm.	Saturated Fat
0.0 gm.	Fiber
1.2 gm.	Carbohydrate
211.4 mg.	Sodium
100 mg.	Calcium

Goat-Cheese Spread

Serves 6

Goat cheese has become one of the favorite cheeses in fashioning new recipes from pizzas to salads. Here we are taking some of the fat calories out while keeping all the wonderful flavor.

4 ounces goat cheese

8 ounces light cream cheese

1 teaspoon Vegetable Blend*

1. Mix all ingredients into a food processor with a steel blade until smooth.
2. Place into an airtight container and refrigerate.

Nutritional Analysis (per serving)

160	Calories
12.2 gm.	Fat
38.6 mg.	Cholesterol
68%	of Calories from Fat
9.1 gm.	Protein
7.72 gm.	Saturated Fat
0.07 gm.	Fiber
3.3 gm.	Carbohydrate
230.8 mg.	Sodium
199 mg.	Calcium

*See the herb blends on page 40.

Vegetable Cream-Cheese Spread

Serves 6

Here's a twist on plain cream cheese to fill celery sticks. It's also good with crackers of any kind.

1 package (8 ounces) light cream cheese
1 celery stick, chopped
1 carrot, chopped
½ teaspoon celery seed
¼ teaspoon dillweed
¼ teaspoon thyme
¼ teaspoon marjoram
2 teaspoons parsley

1. Place all ingredients into a food processor with a steel blade. Pulse until well blended and celery and carrots are in tiny chunks.
2. Put the mixture into an airtight container and refrigerate.

Nutritional Analysis (per serving)

94	Calories
6.7 gm.	Fat
20.9 mg.	Cholesterol
64%	of Calories from Fat
4.2 gm.	Protein
4.14 gm.	Saturated Fat
0.54 gm.	Fiber
4.3 gm.	Carbohydrate
121.4 mg.	Sodium
54 mg.	Calcium

Eggplant Caviar

Serves 8

This is an excellent, versatile dish to keep on hand in the refrigerator. In fact, Eggplant Caviar is best if it is refrigerated overnight. It is very low in calories and can be served as a dip with crackers at cocktail time, stuffed in pita bread for lunch, or served on lettuce leaves and garnished with chopped tomatoes and cucumbers as a first course at dinner.

2 medium eggplants
1 tablespoon garlic olive oil
1 small onion, diced
1 red bell pepper, diced
1 tablespoon chopped fresh parsley
½ teaspoon freshly ground black pepper
1 tablespoon fresh lemon juice
3 teaspoons capers

1. Preheat oven to 400°F.
2. Bake whole eggplants for 1 hour. Let cool until you can handle them.
3. Heat olive oil in a skillet; sauté onion and red bell pepper until soft.
4. Cut eggplants in half and scoop out the pulp. Put the pulp in a food processor with a steel blade. Add the onion mixture and pulse 4 times.
5. Add parsley, black pepper, lemon juice, and capers. Pulse 2 to 4 times, until the mixture is well mixed.
6. Place into a covered container and refrigerate at least 6 hours before serving.

Nutritional Analysis (per serving)

28	Calories
1.9 gm.	Fat
0.0 mg.	Cholesterol
60%	of Calories from Fat
0.4 gm.	Protein
0.25 gm.	Saturated Fat
0.84 gm.	Fiber
3 gm.	Carbohydrate
45.4 mg.	Sodium
5 mg.	Calcium

Guacamole

Serves 6

Here is an all-time favorite made with nonfat cottage cheese, so you are getting protein and calcium while you're cutting the fat calories in half! I have made this recipe often in cooking classes and the response is always the same—people who normally don't care for guacamole love it! One of the secrets is to whip the cottage cheese until it is very smooth before adding the avocados. Serve it with baked tortilla chips.

2 cups nonfat cottage cheese
2 ripe avocados, peeled and pitted
Salsa to taste
Fresh cilantro to taste
Baked tortilla chips

1. Place cottage cheese into a food processor with a steel blade and process until smooth.
2. Add avocados and process until desired consistency.
3. Add salsa and cilantro and pulse 2 to 3 times.

Nutritional Analysis
(per serving)

162	Calories
11 gm.	Fat
3.3 mg.	Cholesterol
61%	of Calories from Fat
10.7 gm.	Protein
2.12 gm.	Saturated Fat
1.68 gm.	Fiber
7 gm.	Carbohydrate
312.6 mg.	Sodium
162 mg.	Calcium

Hummus

Serves 10

Hummus is an excellent source of protein and can be very versatile as a dip with crackers and vegetables or as a sandwich spread.

1 cup dried chickpeas

3 tablespoons fresh lemon juice

1 teaspoon roasted garlic

½ cup Tahini butter

½ teaspoon cayenne pepper

Chopped fresh parsley to garnish

1. Soak chickpeas overnight, then rinse.
2. In a large saucepan, cover chickpeas with water and bring to a boil. Cover and simmer 1 hour or until tender.
3. Drain cooked chickpeas, reserving 1 cup liquid.
4. Add remaining ingredients, except parsley, to food processor and process until smooth.
5. Put mixture into a covered bowl and refrigerate at least 6 hours before serving. Garnish with chopped parsley.

Nutritional Analysis (per serving)

156	Calories
6.7 gm.	Fat
0.0 mg.	Cholesterol
39%	of Calories from Fat
8.7 gm.	Protein
1.29 gm.	Saturated Fat
4.57 gm.	Fiber
17.4 gm.	Carbohydrate
64.1 mg.	Sodium
36 mg.	Calcium

Chicken-Liver Pâté

_____
Serves 16
_____

This is almost like a pâté I used to make that was very high in fat. A few modifications to reduce some of the fat and cholesterol and it still tastes great! Serve with water crackers or Melba toast.

2 tablespoons butter
4 green onions, including tops, chopped
1½ pounds chicken livers
½ teaspoon salt
2 teaspoons dry mustard
½ teaspoon ground nutmeg
½ teaspoon ground cloves
16 ounces light cream cheese
¼ cup cognac

1. In a large skillet, melt the butter. Add onions and sauté until tender.
2. Add chicken livers, salt, mustard, nutmeg, and cloves. Cover and cook over low heat for 10 to 15 minutes or until the livers are well cooked.
3. Place into a food processor with a steel blade and process until smooth.
4. While the processor is running, add the cream cheese and then the cognac.
5. Pour the mixture into a pâté tureen or a soufflé dish. Chill for 12 to 24 hours before serving.

Nutritional Analysis
(per serving)

142	Calories
8.7 gm.	Fat
148.3 mg.	Cholesterol
55%	of Calories from Fat
8.7 gm.	Protein
4.67 gm.	Saturated Fat
0.27 gm.	Fiber
5.1 gm.	Carbohydrate
268 mg.	Sodium
45 mg.	Calcium

Country Pâté without Liver

Serves 10

For the times you want to serve a pâté but aren't sure if everyone likes liver, this pâté is just right. It is excellent served with mustard and a spicy tomato sauce. Make it a day ahead, or freeze it to have on hand when you need it. It's a great appetizer for company.

4 chicken breasts, skinned, boned, and ground
4 chicken thighs, boned and ground
½ pound pork, trimmed of all fat and ground
1 medium onion, chopped
2 tablespoons chopped fresh parsley
1 teaspoon fresh ground black pepper
½ teaspoon ground ginger
¼ teaspoon ground cloves
¼ teaspoon cinnamon
1 teaspoon Worcestershire sauce
¼ teaspoon cayenne pepper
1 tablespoon cognac
2 tablespoons sherry or Madeira

1. Preheat oven to 350°F.
2. Combine all ingredients well.
3. Spray a nonstick loaf pan with nonstick spray and press mixture into it. Cover with foil and bake for one hour.

Nutritional Analysis (per serving)

164	Calories
4.1 gm.	Fat
73.7 mg.	Cholesterol
23%	of Calories from Fat
26.8 gm.	Protein
1.26 gm.	Saturated Fat
0.32 gm.	Fiber
1.7 gm.	Carbohydrate
68.6 mg.	Sodium
22 mg.	Calcium

Spinach and Chicken Pâté

―――――――――
Serves 6
―――――――――

This pâté makes an excellent appetizer or main course for a luncheon or light supper. Serve with a big green salad and a good fresh bread.

1 shallot

1 clove garlic

1 package (10 ounces) frozen spinach, thawed, with water squeezed out

1 pound chicken breast, boned, skinned, and chopped

1 tablespoon arrowroot

1 cup 1% milk

¼ teaspoon nutmeg

1 tablespoon fresh tarragon (or ½ tablespoon dried)

½ teaspoon cayenne pepper

½ teaspoon freshly ground black pepper

¼ teaspoon salt

8 ounces egg substitute

1. Preheat oven to 375°F.
2. In a food processor with a steel blade, add shallots and garlic while the processor is running and mince well.
3. Add remaining ingredients to the processor and pulse on and off until all is mixed well.
4. Put mixture into a loaf pan and cover with foil.
5. Bake for 45 minutes. Let cool 10 minutes before unmolding.

Nutritional Analysis
(per serving)

190	Calories
4.7 gm.	Fat
66 mg.	Cholesterol
22%	of Calories from Fat
31.2 gm.	Protein
1.34 gm.	Saturated Fat
1.22 gm.	Fiber
4.8 gm.	Carbohydrate
437.5 mg.	Sodium
162 mg.	Calcium

Artichokes with Shrimp

___

Serves 4
___

This dish presents exceptionally well. It is a great first course or a main dish for a lunch or light summer supper.

4 artichokes
1 lemon
8 ounces cooked shrimp, cubed
2 celery stalks, diced
½ cup light sour cream
¼ cup low-calorie mayonnaise
1 tablespoon dillweed
⅛ teaspoon garlic powder
4 large red lettuce leaves
1 can (4 ounces) mandarin oranges, drained

1. Cut the tops off artichokes, remove the small leaves, trim the points off the remaining leaves, and trim the stem to make the bottom flat. Cut lemon in fourths. Rub lemon on all cut portions of the artichokes.
2. In a large pot cover half of the artichokes with water. Add the lemon quarters to the water. Cover and simmer for 30 to 45 min or until the center is tender.
3. Drain the artichokes upside down on paper towels. Discard the lemons. When the artichokes cool, remove the center and scrape out the fuzz.
4. Mix shrimp, celery, sour cream, mayonnaise, dillweed, and garlic powder. Chill for 1 hour.
5. Fill artichokes with shrimp mixture and place on individual serving plates lined with red lettuce. Garnish with mandarin oranges, putting some oranges between the leaves of the artichokes.

Nutritional Analysis (per serving)

217	Calories
7.5 gm.	Fat
107 mg.	Cholesterol
31%	of Calories from Fat
18.3 gm.	Protein
2.66 gm.	Saturated Fat
7.32 gm.	Fiber
22.7 gm.	Carbohydrate
321 mg.	Sodium
126 mg.	Calcium

Chicken Drumettes

___

Serves 6
___

These are wonderful finger foods for a picnic or appetizers. In case you are wondering what drumettes are, they are the part of the chicken wing that looks like a baby drumstick! You can use the recipe with other parts of the chicken as a main course.

30 chicken drumettes
1 cup light soy sauce
2 tablespoons sugar
1½ teaspoons ground ginger
½ teaspoon five-spice powder
4 green onions, thinly sliced

1. Blanch chicken drumettes in boiling water for 5 minutes. Rinse in cold water and remove skins.
2. Mix all ingredients, except chicken.
3. Add chicken to marinade, cover, and refrigerate overnight.
4. Preheat oven to 350°F.
5. Place chicken on baking sheet that has been sprayed with nonstick spray and bake for 45 minutes. Serve hot or cold.

Nutritional Analysis
(per serving)

311	Calories
13.7 gm.	Fat
110.8 mg.	Cholesterol
40%	of Calories from Fat
35.8 gm.	Protein
3.75 gm.	Saturated Fat
0.79 gm.	Fiber
9.4 gm.	Carbohydrate
2856.4 mg.	Sodium
37 mg.	Calcium

Crab Cakes

Serves 6

These crab cakes can be served as an appetizer, or on top of a salad as a main course.

1 pound fresh crabmeat (or imitation if fresh is not available)
½ cup bread crumbs
½ teaspoon Basic Herb Blend*
2 egg whites
2 tablespoons low-calorie mayonnaise
2 tablespoons 1% milk
2 tablespoons green onion, minced
1 tablespoon chopped fresh parsley
1 teaspoon Worcestershire sauce
1 teaspoon Dijon mustard
Dash Tabasco sauce
¼ teaspoon ground white pepper
⅔ cup flour
½ teaspoon paprika
⅛ teaspoon cayenne pepper (or to taste)
2 lemons, cut in wedges

Nutritional Analysis (per serving)

171	Calories
2.9 gm.	Fat
77.1 mg.	Cholesterol
15%	of Calories from Fat
18.9 gm.	Protein
0.47 gm.	Saturated Fat
0.99 gm.	Fiber
16.6 gm.	Carbohydrate
326.9 mg.	Sodium
105 mg.	Calcium

1. In a large bowl, combine crabmeat, bread crumbs, Basic Herb Blend, egg whites, mayonnaise, milk, onion, parsley, Worcestershire sauce, mustard, Tabasco, and white pepper.
2. Mix flour, paprika, and cayenne pepper together.
3. Divide crab mixture into 12 balls. Flatten to make cakes and then dip each in flour mixture to coat.
4. In a large nonstick skillet sprayed with nonstick spray, fry each cake 2 minutes on each side.
5. Serve with lemon wedges.

*See the herb blends on page 40.

Phyllo Goat-Cheese Squares

Serves 6

This makes an excellent first course. It can be a bit tricky to make; the secret is to keep the phyllo dough from drying out. Keep the dough covered with a damp cloth while working. These come out in squares.

6 sheets phyllo dough
1 recipe Goat-Cheese Spread (page 46)
1 recipe Tricolor Peppers (page 188)

1. Take 1 sheet of phyllo dough at a time; spray it lightly with olive-oil spray. Fold it over and spray lightly again. Fold again. Place 2 ounces of Goat-Cheese Spread in the center and fold over, spraying lightly again with olive oil. Fold over once more and spray again.
2. Repeating step 1, make a total of 6 squares. Place on baking sheet sprayed with olive oil and bake at 350°F for 30 minutes in preheated oven.
3. Make Tricolor Peppers.
4. Place each phyllo square on a salad plate and spoon peppers alongside.

Nutritional Analysis (per serving)

193	Calories
12.4 gm.	Fat
38.6 mg.	Cholesterol
58%	of Calories from Fat
10.3 gm.	Protein
7.77 gm.	Saturated Fat
1.2 gm.	Fiber
10.4 gm.	Carbohydrate
297.9 mg.	Sodium
212 mg.	Calcium

Salads

Making a Great Salad

Bibb Lettuce with Radishes
Cabbage-Carrot Coleslaw
Cucumber and Tomatoes
Dilled Shrimp Salad
Fruit Salad with Dressing
Greek Salad
Green Beans Vinaigrette
Rice Salad Oriental
Romaine, Pears, and Walnuts
Spinach and Feta Salad
Spinach and Red Cabbage Salad
Spinach, Raspberry, and Walnut Salad
Tomato Basil with Fresh Mozzarella
Waldorf Salad

Making a Great Salad

Salads can be an adventure, with many unique combinations that are both satisfying and healthful. The component that normally takes the salad into the high-calorie and high-fat range is the dressing. Changing how you create your salads can alter that by balancing fats and making the best use of them. One way is to dress your salad with seasoned rice wine vinegar or an herbed vinegar. Most vegetables are low in calories and high in fiber. They can be full of flavor and are a great way to satisfy hunger. In fact, the following vegetables are on the free-food list, which means they contain fewer than 20 calories per serving:

- Cabbage
- Celery
- Chinese cabbage
- Cucumber
- Green onion
- Herbs (fresh)
- Hot peppers
- Mushrooms
- Radishes
- Zucchini
- Endive
- Escarole
- Lettuce
- Romaine
- Spinach

If you wash the greens and spin them dry in a salad spinner, they will hold well in the refrigerator for four to five days. This gives you a ready base to create dozens of different salads, depending on what you have on hand. A diced pear or an apple gives a fresh taste to a salad; a couple of tablespoons of chopped nuts gives it a whole new flavor. You can easily balance out your salads by taking note of the items that are high in fat and adding only one or two. To make a salad into a main dish, you can add a little extra protein, such as legumes, cottage cheese, diced meat or chicken, shrimp, crabmeat, or tuna.

Here is a list of items you can add to salads according to their caloric value.

25 calories
- ½ cup cooked asparagus
- 1 cup raw bean sprouts
- 1 cup raw carrots
- 1 cup green, yellow, or red bell peppers
- 1 large tomato
- ½ cup cooked beets
- 1 cup raw beets

55 calories
- 1 ounce low-fat cheese, including Parmesan

60 calories
- 1 apple
- 1 pear
- 2 tablespoons raisins
- ½ cup mandarin oranges

70 calories
- ½ cup nonfat cottage cheese

80 calories

- ⅓ cup beans (black, navy, pinto, kidney)
- ½ cup corn
- ½ cup lima beans
- 3 tablespoons Grape Nuts
- ½ cup peas

45 fat calories

- 10 small olives
- 2 whole walnuts
- 6 dry roasted almonds
- 1 tablespoon pine nuts
- 1 tablespoon shelled sunflower seeds
- ⅛ medium avocado
- 2 tablespoons shredded coconut
- 1 tablespoon most regular salad dressings

Take the time to experiment. Your favorite combinations will become the signature salads that your family and friends will love.

Bibb Lettuce with Radishes

___
Serves 6
___

Sometimes simple is better and you cannot get much simpler than this.

1 head bibb lettuce, washed, dried, and torn into small pieces

12 radishes, thinly sliced

DRESSING

1 tablespoon olive oil

3 tablespoons herb wine vinegar

½ teaspoon Vegetable Blend*

2 tablespoons water

1 tablespoon Dijon mustard

⅛ teaspoon freshly ground black pepper

1. Put lettuce and radishes in salad bowl.
2. Mix ingredients of dressing together and toss with salad.

*See the herb blends on page 40.

Nutritional Analysis (per serving)

27	Calories
2.5 gm.	Fat
0.0 mg.	Cholesterol
85%	of Calories from Fat
0.3 gm.	Protein
0.33 gm.	Saturated Fat
0.42 gm.	Fiber
1.3 gm.	Carbohydrate
34.5 mg.	Sodium
9 mg.	Calcium

Cabbage-Carrot Coleslaw

Serves 6

This coleslaw adds great color along with great flavor. You can buy the cabbage and carrots preshredded in most grocery stores.

3 tablespoons reduced-fat mayonnaise

3 tablespoons light sour cream

1 tablespoon Dijon mustard

2 tablespoons herbed wine vinegar or seasoned rice wine vinegar

1 teaspoon sugar

½ teaspoon caraway seed

Salt and freshly ground black pepper to taste

2 cups shredded red cabbage

2 cups shredded green cabbage

1 cup grated carrots

1. In large bowl, combine mayonnaise, sour cream, mustard, vinegar, sugar, and caraway seed.
2. Add salt and pepper.
3. Add cabbage and carrots and toss well.
4. Serve right away or chill and serve within 2 hours.

Nutritional Analysis (per serving)

57	Calories
3.3 gm.	Fat
5 mg.	Cholesterol
51%	of Calories from Fat
1.1 gm.	Protein
1.2 gm.	Saturated Fat
2.13 gm.	Fiber
6.7 gm.	Carbohydrate
88.3 mg.	Sodium
37 mg.	Calcium

Cucumber and Tomatoes in Lettuce Cups

Serves 6

A great salad that will go with just about anything—simple yet elegant!

6 lettuce-leaf cups
2 medium-ripe tomatoes, diced
1 English cucumber, diced
½ teaspoon Vegetable Blend*
¼ teaspoon salt
2 tablespoons seasoned rice wine vinegar

1. Place lettuce cups on 6 plates.
2. Mix remaining ingredients and place in lettuce cups.
3. Serve or hold in refrigerator until ready to serve.

Nutritional Analysis (per serving)

16   Calories
0.2 gm.   Fat
0.0 mg.   Cholesterol
13%   of Calories from Fat
0.7 gm.   Protein
0.04 gm.   Saturated Fat
0.82 gm.   Fiber
3.7 gm.   Carbohydrate
86 mg.   Sodium
11 mg.   Calcium

---
*See the herb blends on page 40.

Dilled Shrimp Salad

Serves 6

The perfect first course for six or the main course for lunch to serve four. Rinsing the shrimp gives them a cleaner, fresher flavor. You can also give them a dunk in some ice water that has fresh lemon juice added.

1 pound shrimp, cleaned and deveined
1 can (5 ounces) sliced water chestnuts
3 tomatoes, diced
4 tablespoons seasoned rice wine vinegar
2 tablespoons fresh dillweed, minced
6 romaine leaves, trimmed, rinsed, and patted dry
Fresh parsley to garnish

1. Rinse shrimp in cold water, drain, and pat dry.
2. Toss shrimp with water chestnuts, tomatoes, vinegar, and dillweed.
3. Chill well.
4. Place shrimp salad on romaine leaves and garnish with fresh parsley.

Nutritional Analysis (per serving)

115	Calories
1.7 gm.	Fat
113.3 mg.	Cholesterol
13%	of Calories from Fat
16.9 gm.	Protein
0.30 gm.	Saturated Fat
2.59 gm.	Fiber
8.7 gm.	Carbohydrate
117 mg.	Sodium
96 mg.	Calcium

Fruit Salad with Dressing

Serves 6

You can use a variety of fruits for this salad, including canned fruits in light syrup or frozen. Depending on the time of year, canned fruit may be a better choice. Be sure to buy good-quality fruit.

DRESSING

$2/3$ cup white wine vinegar

$2/3$ cup fruit nectar

$1/4$ teaspoon marjoram

$1/2$ teaspoon ground nutmeg

$1/2$ teaspoon orange peel

$1/4$ teaspoon lemon peel

2 teaspoons chervil

2 apples, cored and sliced

3 apricots, quartered

1 cantaloupe, cut in pieces

2 cups strawberries, hulled

1 head iceberg lettuce, cut into cubes

1. Place all the ingredients for the dressing into a container and shake well.
2. Mix the apples, apricots, cantaloupe, strawberries, and lettuce in a large bowl.
3. Pour the salad dressing over the fruit and lettuce.
4. Toss well and serve.

Nutritional Analysis (per serving)

74	Calories
0.5 gm.	Fat
0.0 mg.	Cholesterol
7%	of Calories from Fat
1 gm.	Protein
0.1 gm.	Saturated Fat
2.99 gm.	Fiber
18.7 gm.	Carbohydrate
4.8 mg.	Sodium
24 mg.	Calcium

Greek Salad

Serves 6

The flavors in this Greek Salad marry wonderfully, and it looks as good as it tastes!

1 head Boston lettuce
1 medium English cucumber, sliced
2 medium tomatoes, cut in wedges
½ medium red onion, sliced
2 medium green bell peppers, sliced
1 tablespoon minced fresh oregano (1 teaspoon dried)
3 tablespoons wine vinegar
½ teaspoon black pepper
4 ounces feta cheese, crumbled

1. Wash and dry lettuce, keeping the leaves whole. Line a platter or shallow bowl with the leaves.
2. Mix cucumber, tomato wedges, and onion and bell pepper slices together.
3. Mix oregano, vinegar, and pepper together, and toss gently with the vegetable mixture.
4. Place vegetables on top of the lettuce leaves and sprinkle with feta cheese.

Nutritional Analysis
(per serving)

78	Calories
4.3 gm.	Fat
16.6 mg.	Cholesterol
50%	of Calories from Fat
3.8 gm.	Protein
2.85 gm.	Saturated Fat
1.43 gm.	Fiber
7.3 gm.	Carbohydrate
214.2 mg.	Sodium
116 mg.	Calcium

Green Beans Vinaigrette

___
Serves 6
___

Green Beans Vinaigrette doubles as a salad and as the vegetable. It's great with pasta.

1 tablespoon Dijon mustard
3 tablespoons herb vinegar
2 tablespoons olive oil
¼ cup water
1 teaspoon fresh lemon juice
½ teaspoon Italian Blend*
½ teaspoon sugar
½ teaspoon freshly ground black pepper
1 ½ pound fresh green beans, trimmed
6 large lettuce cups

1. Combine everything except the green beans and the lettuce cups and set aside.
2. Steam green beans until tender, place into bowl, and pour the vinaigrette over them.
3. Mix well and refrigerate for at least 2 hours.
4. Serve in lettuce cups.

*See the herb blends on page 40.

Nutritional Analysis
(per serving)

87	Calories
5.1 gm.	Fat
0.0 mg.	Cholesterol
53%	of Calories from Fat
2.4 gm.	Protein
0.71 gm.	Saturated Fat
1.68 gm.	Fiber
10.3 gm.	Carbohydrate
36.6 mg.	Sodium
62 mg.	Calcium

Rice Salad Oriental

Serves 6

With this dish you get three in one—a salad, a vegetable, and a starch! It makes a great summer side dish. Add grilled shrimp or chicken, and you have the whole meal.

2 cups brown rice, cooked
1 tablespoon instant onion, toasted
½ cup radishes, sliced
½ green pepper, sliced
2 small celery stalks, sliced
1 tablespoon low-sodium soy sauce
1 tablespoon chopped fresh cilantro
¼ teaspoon ground ginger
⅛ teaspoon garlic powder
¼ cup seasoned rice wine vinegar
1 head red leaf lettuce, keep out 6 leaves and chop the rest

1. Mix all the ingredients, except the lettuce, in a large bowl. Toss well to mix. Cover and refrigerate for 1 hour.
2. Toss in the chopped lettuce. Line a salad bowl with remaining lettuce leaves and place the mixture in the center.

Nutritional Analysis (per serving)

87	Calories
0.5 gm.	Fat
0.0 mg.	Cholesterol
5%	of Calories from Fat
2.1 gm.	Protein
0.11 gm.	Saturated Fat
1.67 gm.	Fiber
18.6 gm.	Carbohydrate
180.6 mg.	Sodium
19 mg.	Calcium

Romaine, Pears, and Walnuts

Serves 4

A fast salad that is also very refreshing, it works well as a palate cleaner if you have served appetizers before dinner.

1 head romaine lettuce, washed and cut in bite-size pieces
2 pears, cut in cubes and tossed in lemon water
¼ cup walnuts, chopped
¼ cup fruit-infused vinegar

Toss all ingredients together and serve.

Nutritional Analysis (per serving)

102	Calories
5.2 gm.	Fat
0.0 mg.	Cholesterol
46%	of Calories from Fat
1.7 gm.	Protein
0.46 gm.	Saturated Fat
2.87 gm.	Fiber
14.5 gm.	Carbohydrate
4.3 mg.	Sodium
25 mg.	Calcium

Spinach and Feta

Serves 6

Check the date on the bag of spinach to make sure it is fresh.

1 pound fresh spinach, washed, dried, and torn into small pieces
¼ cup diced walnuts
¼ cup wine vinegar
¼ cup water
2 tablespoons chopped fresh basil
4 ounces feta cheese, crumbled
½ teaspoon freshly ground black pepper

1. Place spinach and walnuts into a large salad bowl.
2. Mix the remaining ingredients together.
3. Pour dressing on spinach and toss well to coat.

Nutritional Analysis (per serving)

104	Calories
7.5 gm.	Fat
16.6 mg.	Cholesterol
65%	of Calories from Fat
5.7 gm.	Protein
3.12 gm.	Saturated Fat
2.42 gm.	Fiber
5.7 gm.	Carbohydrate
269.9 mg.	Sodium
198 mg.	Calcium

Spinach and Red Cabbage Salad with Honey-Mustard Dressing

Serves 6

This salad has eye appeal along with great flavor. It is very good served with Mustard Chicken (see page 147).

1 bag baby spinach

1 bag red cabbage, shredded

½ cup white wine vinegar

2 tablespoons country mustard with seeds

¼ cup honey

1. Rinse and spin-dry spinach and cabbage.
2. Mix vinegar, mustard, and honey well, pour over spinach and cabbage, and toss.

Nutritional Analysis (per serving)

64	Calories
0.4 gm.	Fat
0.0 mg.	Cholesterol
5%	of Calories from Fat
1.5 gm.	Protein
0.04 gm.	Saturated Fat
2.09 gm.	Fiber
16 gm.	Carbohydrate
91.8 mg.	Sodium
50 mg.	Calcium

Spinach, Raspberry, and Walnut Salad

Serves 6

If raspberries are not in season, use canned or frozen. If you have a hard time finding raspberry juice, substitute any berry or fruit nectar.

1½ pounds fresh spinach, washed and dried
1 cup fresh raspberries
¼ cup chopped walnuts
1 tablespoon chopped fresh tarragon (1 teaspoon dried)
¼ cup raspberry vinegar
½ cup raspberry juice

1. Tear spinach into small pieces.
2. In a salad bowl, place spinach, raspberries, walnuts, and tarragon.
3. Mix together vinegar and raspberry juice. Pour over salad and toss to coat.

Nutritional Analysis (per serving)

78	Calories
3.9 gm.	Fat
0.0 mg.	Cholesterol
44%	of Calories from Fat
4.5 gm.	Protein
0.36 gm.	Saturated Fat
4.78 gm.	Fiber
9.6 gm.	Carbohydrate
90.9 mg.	Sodium
136 mg.	Calcium

Tomato Basil with Fresh Mozzarella and Balsamic Vinegar

Serves 4

This can also be served by slicing the tomatoes and cheese and alternating them on salad plates. Tuck the whole basil leaves underneath and drizzle the balsamic vinegar on top.

2 Roma tomatoes, seeded and chopped
8 ounces fresh mozzarella packed in water, cubed
Fresh basil leaves, thinly sliced
Balsamic vinegar to taste
Lettuce cups

Mix all ingredients together and serve in lettuce cups.

Nutritional Analysis (per serving)

197	Calories
13.1 gm.	Fat
39.2 mg.	Cholesterol
60%	of Calories from Fat
15.9 gm.	Protein
8.27 gm.	Saturated Fat
0.8 gm.	Fiber
4.6 gm.	Carbohydrate
301.2 mg.	Sodium
413 mg.	Calcium

Waldorf Salad

Serves 4

The Waldorf is always a favorite salad, any time of the year. I like dates in this salad to make it a bit different. Toss chopped apples in lemon water to keep them from browning.

2 apples, cored and chopped into bite-size pieces
2 stalks celery, sliced diagonally
½ cup date pieces, or raisins
¼ cup chopped walnuts
½ cup light sour cream
2 tablespoons light mayonnaise
⅛ teaspoon nutmeg

1. Combine apples, celery, dates, and walnuts.
2. Mix sour cream with mayonnaise and nutmeg. Toss with apple mixture to coat.
3. Cover and chill.

Nutritional Analysis (per serving)

221	Calories
9.8 gm.	Fat
8.3 mg.	Cholesterol
40%	of Calories from Fat
2.4 gm.	Protein
2.69 gm.	Saturated Fat
4.6 gm.	Fiber
35.2 gm.	Carbohydrate
66.4 mg.	Sodium
46 mg.	Calcium

Soups

They Warm the Soul!

Corn
Gazpacho
Golden Squash
Mushroom and Barley
Tomato-Herb Bouillon
Yogurt-Cucumber

They Warm the Soul!

Soup can warm you up on a cold winter evening, and it's a meal in itself if you add some good crusty bread topped with a cheese spread. On warm summer nights a cold Gazpacho or Yogurt-Cucumber soup can be the ticket. As a first course, Tomato Herb Bouillon is light and adds very few calories to the meal.

To give canned soup a kick, add some chopped fresh herbs right before serving. Bring canned chicken broth to life by adding sliced carrots and celery and a sprig of parsley. Simmer for 20 minutes and strain before using (you can munch on the carrot and celery).

A good choice for a hearty soup is Navy Bean, found on page 91 in the bean section.

Corn Soup

Serves 6

An interesting change of pace as the first-course soup for a dinner party or a luncheon.

½ medium onion, chopped
6 cups low-sodium chicken broth
2 cups corn kernels
1 teaspoon ground ginger
½ teaspoon Basic Blend*
¼ teaspoon salt
¼ teaspoon freshly ground pepper
2 tablespoons low-sodium soy sauce
2 tablespoons water
1 tablespoon cornstarch
4 egg whites
2 green onions, thinly sliced

1. In a large Dutch oven sprayed with olive oil, add onions and sauté until soft.
2. Add chicken broth and bring to a boil. Stir in corn, ginger, Basic Blend, salt, and pepper. Cover, reduce heat, and simmer until hot.
3. Combine soy sauce, water, and cornstarch. Slowly add to soup, stirring constantly until soup is thickened. (If you are making the soup ahead of time, hold the preparation at this point.)
4. Beat the egg whites with the green onion, and gradually add the mixture to the soup, stirring just until the eggs are set.
5. Serve immediately.

*See the herb blends on page 40.

Nutritional Analysis (per serving)

103	Calories
1.3 gm.	Fat
0.0 mg.	Cholesterol
11%	of Calories from Fat
7.5 gm.	Protein
0.39 gm.	Saturated Fat
2.67 gm.	Fiber
17.3 gm.	Carbohydrate
1240.1 mg.	Sodium
29 mg.	Calcium

Gazpacho

Serves 6

You can also turn this Gazpacho into a fantastic Bloody Mary to serve before brunch. Process the whole batch and serve it poured over ice with a stalk of celery for garnish—with or without vodka. Or you could mix it with beer—regular, light, or nonalcoholic!

1 cup cucumber, diced
1 green bell pepper, seeded and diced
1 can (4 ounces) diced green chilies, drained
1 can (46 ounces) tomato juice
½ teaspoon Worcestershire sauce
1½ cup diced tomatoes
1 tablespoon chopped fresh cilantro
1 tablespoon chopped fresh oregano
2 green onions, diced
⅛ teaspoon cayenne pepper (or to taste)

1. Combine all ingredients. Put half of the mixture into a food processor and blend until smooth.
2. Combine the two mixtures in a large bowl. Cover and refrigerate at least 1 hour before serving.

Nutritional Analysis (per serving)

74	Calories
0.5 gm.	Fat
0.0 mg.	Cholesterol
6%	of Calories from Fat
3.1 gm.	Protein
0.08 gm.	Saturated Fat
3.91 gm.	Fiber
17.6 gm.	Carbohydrate
36.8 mg.	Sodium
57 mg.	Calcium

Golden Squash Soup

___

Serves 6

___

Here's a soup that's pleasing not only to the palate, but also to the eye. It may take a little longer to prepare than some of the other recipes, but, believe me, it's worth the extra time.

3 onions, chopped
1 cup celery, chopped
1 clove garlic, minced
½ teaspoon rosemary
¼ teaspoon freshly ground black pepper
1 quart low-sodium chicken broth
2 cups acorn or butternut squash, cooked and mashed
2 cups 1% milk
Nutmeg to garnish

1. Combine onions, celery, garlic, rosemary, and pepper in a soup pot. Spray with olive oil and sauté until onions and celery are soft.
2. Add chicken broth and squash, and heat through.
3. Remove from heat, add milk, and sprinkle with nutmeg.

Nutritional Analysis
(per serving)

102	Calories
1.9 gm.	Fat
3.3 mg.	Cholesterol
16%	of Calories from Fat
6.1 gm.	Protein
0.84 gm.	Saturated Fat
3.46 gm.	Fiber
16.7 gm.	Carbohydrate
582.8 mg.	Sodium
143 mg.	Calcium

Mushroom and Barley Soup

―――――――――
Serves 4
―――――――――

Turn this into a vegetarian soup by using vegetable broth instead of the beef broth. When you add raw mushrooms to the cooking liquid, they stay plumper and give a good texture to the soup, which makes it heartier without extra calories. Serve this soup with a salad and hot, crusty bread for a winter supper.

1 medium onion, chopped
1 clove garlic, minced
1 teaspoon basil, crushed
6 cups low-salt beef broth
1 cup barley
½ teaspoon Worcestershire sauce
⅛ teaspoon pepper
1 pound mushrooms, sliced
½ cup shredded carrots
2 tablespoons cornstarch mixed with 2 tablespoons water
1 tablespoon chopped fresh parsley

1. In a large Dutch oven sprayed with olive oil, sauté onion and garlic until soft.
2. Add basil and beef broth, bring to a boil.
3. Add barley, Worcestershire sauce, and pepper. Cover and simmer 15 minutes.
4. Add mushrooms and carrots. Cover and simmer 25 minutes.
5. Uncover and add cornstarch mixture and cook until thick. Add parsley and serve.

Nutritional Analysis
(per serving)

279	Calories
2 gm.	Fat
0.0 mg.	Cholesterol
6%	of Calories from Fat
12.5 gm.	Protein
0.6 gm.	Saturated Fat
10.9 gm.	Fiber
55.6 gm.	Carbohydrate
1199.7 mg.	Sodium
66 mg	Calcium

Tomato-Herb Bouillon

Serves 4

Here's a really tasty, fast, and easy starter bouillon that's sure to please. It also makes the perfect low-calorie snack to have midmorning or midafternoon.

1 teaspoon beef-flavored bouillon paste
1 cup hot water
3 cups tomato juice
2 tablespoons chopped fresh parsley
1 tablespoon lemon juice
1 teaspoon low-sodium Worcestershire sauce
½ teaspoon dried whole rosemary
½ teaspoon dried whole thyme
¼ teaspoon pepper
Lemon slices to garnish

1. Dissolve bouillon in hot water.
2. Add tomato juice and the rest of the ingredients, reserving the garnish. Bring to a boil, reduce heat, and simmer for 5 minutes. Or microwave on high in a microwave-safe bowl for 3 to 5 minutes or until the mixture is hot.
3. Ladle bouillon into soup bowls or large mugs and garnish with lemon slices.

Nutritional Analysis (per serving)

44	Calories
0.5 gm.	Fat
0.3 mg.	Cholesterol
9%	of Calories from Fat
2.2 gm.	Protein
0.15 gm.	Saturated Fat
1.58 gm.	Fiber
9.5 gm.	Carbohydrate
855.3 mg.	Sodium
52 mg.	Calcium

Yogurt-Cucumber Soup

Serves 6

Perfect for warm spring or summer days. Serve it with a salad topped with Phyllo Goat-Cheese Squares (page 57).

1 container (32 ounces) plain nonfat yogurt
1 large cucumber, diced
½ cup diced walnuts
1 teaspoon fresh dill (or ½ teaspoon dried dillweed)
½ teaspoon freshly ground white pepper
⅛ teaspoon salt
Fresh dill to garnish (or dried dillweed)

1. Mix all ingredients in a large bowl, cover, and refrigerate for at least 1 hour.

2. Garnish with fresh dill and serve.

Nutritional Analysis (per serving)

164	Calories
6.8 gm.	Fat
2.9 mg.	Cholesterol
37%	of Calories from Fat
11.1 gm.	Protein
0.78 gm.	Saturated Fat
0.7 gm.	Fiber
15.8 gm.	Carbohydrate
165.6 mg.	Sodium
343 mg.	Calcium

Beans

A Note about Beans

Black-and-White Bean Salad
Black Bean Spread
Lentil Stew
Navy Bean and Basil Salad
Navy Bean Soup
Red Beans and Rice Salad

A Note about Beans

Beans are an excellent source of protein and fiber. As a bonus, they have no fat at all! It is always better to cook dried beans, if you have a choice. You can cook a large pot of beans and then freeze portions of 1 or 2 cups in individual containers for later use. That way, you'll always have them available when you need them. As a second choice, keep a can of beans in your pantry; just rinse them in a colander under cold water before you use them to remove the excess sodium.

The best way to cook dried beans is to soak them overnight. Rinse them the next day and put them in a large pot of water. Bring to a boil, rinse again in a colander under cold water, then put the beans back into the pot. Cover the beans with water or broth. I prefer broth because it gives the beans additional flavor. Add a couple of bay leaves and simmer until the beans are tender, approximately 1 to 2 hours. Remove the bay leaves. Now you have great beans without too much of the starch; it's the starch that can cause stomach gas.

Black-and-White Bean Salad

Serves 8

If you're cooking your own beans, cook the black beans and white beans separately so they each maintain their color (black beans bleed). If you're using canned beans, rinse each separately. Part of the appeal of this salad is that the color is so interesting.

3 cups black beans, cooked
3 cups navy beans, cooked
1 yellow bell pepper, chopped
1 can (2 ounces) green chilies, chopped
½ red onion, finely chopped
½ cup chopped fresh cilantro
½ teaspoon freshly ground black pepper
⅛ teaspoon crushed red pepper (or to taste)
¼ cup seasoned rice wine vinegar
1 head iceberg lettuce, finely chopped

Mix all ingredients and toss well. Serve.

Nutritional Analysis (per serving)

169	Calories
0.6 gm.	Fat
0.0 mg.	Cholesterol
3%	of Calories from Fat
10.6 gm.	Protein
0.15 gm.	Saturated Fat
7.22 gm.	Fiber
28.9 gm.	Carbohydrate
212.2 mg.	Sodium
60 mg.	Calcium

Black Bean Spread

Serves 8

Black Bean Spread is delicious on crackers, celery, or bread. My favorite, however, is to spread it on a large flour tortilla, roll it up, refrigerate for 2 hours, then slice. This makes an attractive appetizer and is easy to have on hand for company.

4 cups black beans, cooked
½ cup salsa
⅛ teaspoon cayenne pepper
¼ cup fresh cilantro

1. Put all ingredients, except cilantro, into a food processor with a steel blade. Process until smooth.
2. Add cilantro and pulse processor until cilantro is just mixed in.
3. Place into a covered container and refrigerate.

Nutritional Analysis (per serving)

102	Calories
0.4 gm.	Fat
0.0 mg.	Cholesterol
4%	of Calories from Fat
6.3 gm.	Protein
0.11 gm.	Saturated Fat
4.17 gm.	Fiber
17.1 gm.	Carbohydrate
3.2 mg.	Sodium
32 mg.	Calcium

Lentil Stew

Serves 4

Lentils are quick to cook and a good source of protein and fiber. With a big salad on the side and a loaf of crusty bread, you have a nutritious, tasty meal.

1 cup dry lentils
3½ cups low-sodium chicken broth
1 can (14½ ounces) diced tomatoes
1 potato, cubed
2 carrots, chopped
2 celery stalks, chopped
1 tablespoon basil
½ teaspoon chervil
½ teaspoon onion powder
½ teaspoon garlic powder
⅛ teaspoon crushed red pepper

1. Rinse and drain lentils.
2. In a large saucepan, combine all ingredients. Bring to a boil. Reduce heat and simmer, covered, for 45 minutes or until done.

Nutritional Analysis
(per serving)

142	Calories
1.1 gm.	Fat
0.0 mg.	Cholesterol
7%	of Calories from Fat
8.4 gm.	Protein
0.32 gm.	Saturated Fat
6.65 gm.	Fiber
25.5 gm.	Carbohydrate
869.6 mg.	Sodium
99 mg.	Calcium

Navy Bean and Basil Salad

Serves 6

This salad is good served warm, but I prefer it chilled. Add chopped celery if you want it crunchy. This is one case where it is much better to use fresh basil for appearance as well as flavor.

½ cup low-sodium chicken broth
¼ cup white wine vinegar
3 shallots, chopped
¼ cup fresh basil cut into thin strips
3 cups navy beans, cooked
2 ounces sliced black olives
6 large red lettuce leaves
3 ripe tomatoes, chopped

1. In a saucepan, combine broth, vinegar, and shallots, and bring to a boil. Reduce to simmer.
2. Add basil, beans, olives, and stir well. Cover and turn off the heat. Let stand 10 minutes. If you prefer a cold salad, chill it in the refrigerator.
3. Place lettuce leaves on salad plates, spoon bean mixture onto leaves, then top with chopped tomato.

Nutritional Analysis
(per serving)

138	Calories
1.4 gm.	Fat
0.0 mg.	Cholesterol
9%	of Calories from Fat
8.5 gm.	Protein
0.24 gm.	Saturated Fat
6.02 gm.	Fiber
22.8 gm.	Carbohydrate
395.2 mg.	Sodium
54 mg.	Calcium

Navy Bean Soup

Serves 8

Old-fashioned navy bean soup like this can be served with a big salad and a side of Green Chili Cornbread (page 217) for lunch. It can also be a good early dinner or late-night supper after the football game.

1 medium onion, chopped
¼ pound Canadian bacon, chopped
2 carrots, chopped
2 celery stalks, chopped
1 teaspoon marjoram
1 teaspoon chervil
¼ teaspoon crushed red pepper
3 cups low-sodium chicken broth
1 can (14 ounces) diced tomatoes in juice
5 cups navy beans, cooked

1. In a large Dutch oven sprayed with olive oil, sauté onions until soft. Add Canadian bacon, carrots, and celery. Sauté, stirring constantly for 2 minutes.
2. Add remaining ingredients and simmer for 30 minutes.
3. Put 1 cup of the soup into a food processor or blender and blend until smooth. Add back to soup to thicken.

Nutritional Analysis
(per serving)

185	Calories
1.4 gm.	Fat
3.3 mg.	Cholesterol
7%	of Calories from Fat
12.8 gm.	Protein
0.38 gm.	Saturated Fat
8.19 gm.	Fiber
29.6 gm.	Carbohydrate
752.9 mg.	Sodium
86 mg.	Calcium

Red Beans and Rice Salad

Serves 6

The beans and rice together make a complete protein. Steeping the sun-dried tomatoes in chicken broth reconstitutes them so they are easy to chop. But if you have the sun-dried tomatoes that are packed in oil, just pat the oil off and they are ready to chop.

½ cup low-sodium chicken broth
8 sun-dried tomatoes
3 cups red beans, cooked
3 cups brown rice, cooked
½ cucumber, chopped
1 red bell pepper, chopped
3 celery stalks, chopped
3 scallions, chopped
¼ cup white wine vinegar
2 tablespoons prepared mustard
½ teaspoon thyme
½ teaspoon marjoram
½ teaspoon chervil
⅛ teaspoon cayenne pepper

1. In a saucepan, bring chicken broth to a boil. Add sun-dried tomatoes, cover, turn heat off, and let stand for 10 minutes.
2. Remove tomatoes, rinse under cold water, and chop.
3. Mix broth, tomatoes, and all other ingredients in a serving dish. Cover and refrigerate at least 30 minutes before serving.

Nutritional Analysis (per serving)

269	Calories
1.3 gm.	Fat
0.0 mg.	Cholesterol
4%	of Calories from Fat
12.1 gm.	Protein
0.28 gm.	Saturated Fat
7.97 gm.	Fiber
51.9 gm.	Carbohydrate
432.5 mg.	Sodium
83 mg.	Calcium

Pasta

Always at the Ready in Your Pantry

Basic Marinara Sauce
Colorful Pasta Peppers
Fettuccine Alfredo
Fettuccine with Artichokes and Tomatoes
Linguine with Clam Sauce
Linguine with Herbed Clam Sauce
Macaroni and Cheese
Meat Sauce
Pasta with Shrimp, Zucchini, and Tomatoes
Pasta Salad with Crab and Snow Peas
Pot-Roasted Pasta
Prosciutto and Tomato Pasta
Shrimp Scampi Provincial
Sun-Dried Tomatoes, Rosemary, and Thyme
Tomato-Basil Sauce with Mostaccioli
Vermicelli with Basil and Pine Nuts
Vermicelli with Cream Cheese
   and Herbs

Always at the Ready in Your Pantry

I'm sure there must be some people who don't love pasta, but I don't know them!

What I like best is that if you always have pasta, cans of tomatoes, and the basic herbs in your pantry, you've got it made. For instance, I keep prosciutto in 3-ounce portions in lock-top plastic bags in my freezer to make Prosciutto and Tomato Pasta. Then, I just need diced tomatoes, garlic-infused olive oil, and basic pasta, and I'm set! To make sure you're prepared to whip up a quick pasta dish, keep some canned chopped clams, capers, sliced black olives, and Italian seasoning in your pantry as well. Of course, always keep fresh Parmesan cheese. With these ingredients, you can make a clam sauce or a marinara sauce with black olives. Use your creative ability to clean out your refrigerator and make an everything-but-the-kitchen-sink sauce.

Basic Marinara Sauce

Serves 8

You can make a multitude of dishes using this sauce as a base. Freeze it in containers to pull out when needed.

1 tablespoon garlic olive oil
1 medium onion, sliced
1 teaspoon Italian Blend*
1 can (28 ounces) tomato purée
1 can (28 ounces) whole tomatoes with basil
1 can (28 ounces) tomato sauce
1 can (2 ounces) sliced black olives
½ teaspoon freshly ground black pepper
1 tablespoon chopped fresh parsley

1. In a large Dutch oven heat the olive oil and sauté onion until soft. Add Italian Blend and cook with the onion for 1 minute.
2. Add the rest of the ingredients, cover, and simmer for 3 hours.

Nutritional Analysis (per serving)

101	Calories
3.1 gm.	Fat
0.0 mg.	Cholesterol
27%	of Calories from Fat
3.6 gm.	Protein
0.42 gm.	Saturated Fat
5.31 gm.	Fiber
18.2 gm.	Carbohydrate
865.5 mg.	Sodium
80 mg.	Calcium

*See the herb blends on page 40.

Colorful Pasta Peppers

Serves 8

This is an excellent side dish for a roasted chicken or a Cornish game hem. It provides the vegetable and the starch.

4 large bell peppers, cut in half and seeded (or you can use red or yellow peppers for variety)

2 cups small pasta shells, cooked al dente, cooked just before serving

¼ cup chopped fresh basil

2 garlic cloves, minced

2 tablespoons chopped chives

4 Roma tomatoes, chopped

2 shallots, minced

2 cups grated mozzarella cheese

½ teaspoon salt

¼ teaspoon freshly ground black pepper

1. Preheat oven to 350°F.
2. Place pepper halves in a baking dish sprayed with nonstick spray.
3. Mix remaining ingredients well and fill pepper halves.
4. Bake for 40 minutes.

Nutritional Analysis (per serving)

166	Calories
7 gm.	Fat
19.6 mg.	Cholesterol
38%	of Calories from Fat
10.3 gm.	Protein
4.19 gm.	Saturated Fat
2.03 gm.	Fiber
16.4 gm.	Carbohydrate
275.7 mg.	Sodium
218 mg.	Calcium

Fettuccine Alfredo

Serves 6

Fettuccine Alfredo, with its cheese and cream, is sometimes known as a heart attack on a plate! This variation falls within the guidelines of healthy eating, while tasting exceptionally good! Use the Alfredo sauce as a springboard for your own culinary creativity. Spoon it over cooked chicken and top with sautéed mushrooms or sliced ripe olives. Add julienne vegetables for color and flavor, and you have a primavera sauce. Diced Canadian bacon and baby peas give you a carbonara sauce. The secret to the sauce is a very good quality Parmesan cheese.

1 tablespoon butter

1½ cups 1% milk

2 tablespoons cornstarch

1 cup light sour cream

¼ teaspoon cayenne pepper

Freshly ground white or black pepper to taste

3 ounces Parmesan cheese, freshly grated

1 pound fettuccine pasta, cooked al dente, cooked just before serving

Chopped fresh parsley to garnish

1. In a large saucepan melt butter. Mix milk and cornstarch, and add to pan. Stir until thick.
2. Mix in sour cream, peppers, and Parmesan cheese.

Toss sauce with pasta, garnish with parsley, and serve.

Nutritional Analysis (per serving)

450	Calories
12.4 gm.	Fat
27.4 mg.	Cholesterol
25%	of Calories from Fat
18.6 gm.	Protein
7.04 gm.	Saturated Fat
3.47 gm.	Fiber
65 gm.	Carbohydrate
324.6 mg.	Sodium
306 mg.	Calcium

Fettuccine with Artichokes and Tomatoes

Serves 8

A simple recipe with a trace of the unique for those of you who love artichokes. When time and availability allow, use fresh small artichokes. When using fresh artichokes, be sure to remove the tough outer leaves and steam the artichokes until they are just tender. When they are cool enough to touch, cut in quarters and remove the choke.

Nutritional Analysis (per serving)

330	Calories
3.8 gm.	Fat
5.5 mg.	Cholesterol
10%	of Calories from Fat
14.2 gm.	Protein
1.58 gm.	Saturated Fat
9.52 gm.	Fiber
61.7 gm.	Carbohydrate
314.9 mg.	Sodium
194 mg.	Calcium

1 medium onion, chopped
2 carrots, cut in julienne strips
2 cans (28 ounces) whole or diced tomatoes
2 teaspoons fresh thyme (½ teaspoon dried)
2 teaspoons chopped fresh rosemary (½ teaspoon dried)
¼ teaspoon freshly ground black pepper
2 packages (9 ounces) frozen artichoke hearts, thawed
2 ounces Parmesan cheese, freshly grated
1 pound fettuccine, cooked al dente, cooked just before serving

1. In a large cast-iron pot or Dutch oven sprayed with olive-oil spray, sauté onions until soft.
2. Add carrots, tomatoes with juice (if you are using whole tomatoes, chop them), thyme, rosemary, and pepper. Simmer uncovered 25 minutes.
3. Cut the artichokes in quarters and add them to the sauce. Simmer for 15 minutes, uncovered.
4. Add the fettuccine and Parmesan, and toss.

Linguine with Clam Sauce

Serves 4

These are ingredients that are easy to always have on hand to cook up a last-minute meal. The capers add a wonderful zip!

- 2 tablespoons garlic-infused olive oil
- 2 tablespoons capers
- 1 can (15 ounces) diced tomatoes in juice
- 1 can (16½ ounces) chopped clams
- ½ teaspoon freshly ground black pepper
- 3 tablespoons chopped fresh parsley
- ½ pound linguine, cooked al dente, cooked just before serving

1. Heat olive oil in a large nonstick skillet; sauté capers for 1 minute.
2. Add tomatoes, clams, pepper, and parsley, and simmer for 10 minutes uncovered.
3. Toss with pasta and serve.

Nutritional Analysis (per serving)

340	Calories
8.8 gm.	Fat
13.8 mg.	Cholesterol
23%	of Calories from Fat
13.9 gm.	Protein
1.17 gm.	Saturated Fat
4.9 gm.	Fiber
51.3 gm.	Carbohydrate
218.3 mg.	Sodium
73 mg.	Calcium

Linguine with Herbed Clam Sauce

Serves 6

Another version of Linguine with Clam Sauce, this time with more herbs and garlic. Fresh herbs are much better in this recipe; however, if they are not easily available, dried will work.

1 tablespoon olive oil
1 onion, chopped
4 cloves garlic, chopped
2 carrots, chopped
½ cup red wine (or dry vermouth)
2 cans (28 ounces) diced tomatoes with juice
1 tablespoon chopped fresh oregano (1 teaspoon dried)
1 tablespoon chopped fresh thyme (½ teaspoon dried)
¼ teaspoon freshly ground black pepper
1 can (2 ounces) chopped clams, drained
1 tablespoon chopped fresh parsley
1 pound linguine, cooked al dente, cooked just before serving

1. In a large skillet heat olive oil and sauté onion, garlic, and carrots until soft.
2. Add red wine, bring to a boil. Add tomatoes, oregano, thyme, and pepper. Simmer uncovered for 25 minutes.
3. Add clams and parsley, stirring well. Pour sauce on linguine and toss.

Nutritional Analysis
(per serving)

421	Calories
4.6 gm.	Fat
2.8 mg.	Cholesterol
10%	of Calories from Fat
14.2 gm.	Protein
0.63 gm.	Saturated Fat
9.52 gm.	Fiber
76.9 gm.	Carbohydrate
55.1 mg.	Sodium
112 mg.	Calcium

Macaroni and Cheese

_____

Serves 6

_____

An old family favorite with only a few minor alterations to make it fit within the guidelines of good health. If you are making it ahead and reheating it, place it in a casserole dish, top with bread crumbs and grated cheese, cover, and bake at 350°F for 20 minutes.

2 tablespoons cornstarch
1½ cups 1% milk
½ teaspoon dry mustard
¼ teaspoon ground white pepper
1 teaspoon Worcestershire sauce
½ teaspoon Vegetable Blend*
¼ teaspoon cayenne pepper
2 cups sharp cheddar cheese, grated
1 pound elbow macaroni, cooked al dente
Garnish with freshly chopped parsley and paprika

1. Mix all ingredients together except cheese and macaroni. Heat in a large Dutch oven, stirring constantly to bring mixture to a simmer. Keep stirring until the mixture has thickened.
2. Add cheese and stir until it is melted. Add in cooked pasta, stir gently until mixed well.
3. Serve right away or pour into a baking dish, cover, and keep warm in the oven.
4. Garnish with parsley and paprika.

Nutritional Analysis (per serving)

484	Calories
14.6 gm.	Fat
41.7 mg.	Cholesterol
27%	of Calories from Fat
21.6 gm.	Protein
8.5 gm.	Saturated Fat
3.59 gm.	Fiber
65.3 gm.	Carbohydrate
273.1 mg.	Sodium
364 mg.	Calcium

_____

*See the herb blends on page 40.

Meat Sauce

Serves 8

A versatile meat sauce that you can serve over thick pasta noodles, or mix with macaroni or rice and top with cheese. It can be put into a crock-pot or slow cooker and cooked all day while you are at work or play.

1 tablespoon olive oil
1 medium onion, chopped
1 clove garlic, minced
2 medium carrots, chopped
1 medium green pepper, chopped
1 teaspoon Italian Blend*
2 tablespoons chopped fresh parsley
1 pound extra-lean ground beef
1 can (28 ounces) diced tomatoes in juice
1 can (28 ounces) tomato sauce
1 bay leaf

1. In a large Dutch oven heat olive oil and sauté onion, garlic, carrots, and green pepper until soft.

2. Add Italian Blend and sauté 30 seconds. Add parsley.

3. In a nonstick skillet, sauté ground beef until done. Drain on paper towels to remove excess fat.

4. Add beef to Dutch oven. Add tomatoes and tomato sauce and bring to a simmer. Add bay leaf, cover, and simmer for 2 hours. Remove bay leaf before serving.

Nutritional Analysis (per serving)

175	Calories
6.6 gm.	Fat
38.5 mg.	Cholesterol
34%	of Calories from Fat
14.7 gm.	Protein
1.92 gm.	Saturated Fat
4.52 gm.	Fiber
16.2 gm.	Carbohydrate
693.1 mg.	Sodium
62 mg.	Calcium

*See the herb blends on page 40.

Pasta with Shrimp, Zucchini, and Tomatoes

Serves 6

A light, lovely-to-look-at, fantastic-to-eat luncheon or supper entrée.

4 small zucchini, thinly sliced

¼ teaspoon coarse salt

1 pound shrimp (30 to a pound), cooked, cut in half lengthwise

1 lemon

2 tablespoons garlic-infused olive oil

1 cup chicken broth, warm

Freshly ground black pepper to taste

3 tomatoes, peeled, seeded, and chopped

1 pound fettuccine, cooked al dente, cooked just before serving

Fresh Italian parsley

1. Put sliced zucchini in a bowl of ice water with ¼ teaspoon coarse salt for 30 minutes.
2. Put shrimp in cold water with the juice of the lemon.
3. Heat olive oil in a large skillet and sauté zucchini until soft.
4. Add chicken broth, bring to a boil, and reduce to simmer.
5. Add shrimp and chopped tomatoes. Simmer until well heated, about 3 minutes.
6. Pour over warm fettuccine and toss. Garnish with Italian parsley and serve.

Nutritional Analysis (per serving)

416	Calories
7.4 gm.	Fat
60 mg.	Cholesterol
16%	of Calories from Fat
20 gm.	Protein
1.13 gm.	Saturated Fat
5.92 gm.	Fiber
67.6 gm.	Carbohydrate
188.9 mg.	Sodium
75 mg.	Calcium

Pasta Salad with Crab and Snow Peas

Serves 6

A perfect Sunday lunch pasta salad that will please all.

4 shallots, thinly sliced

1 cup vegetable broth

2 tablespoons herbed wine vinegar

1 pound snow peas, string removed, sliced diagonally

1 tablespoon chopped fresh basil (½ teaspoon dried)

Freshly ground black pepper, to taste

1 pound rotini pasta, cooked al dente, rinsed in cold water

2 tablespoons chopped fresh parsley

1 pound imitation crabmeat

1. Spray a skillet with olive oil and sauté shallot until soft.
2. Add vegetable broth and vinegar. Bring to a boil, and reduce heat to simmer.
3. Add snow peas, basil, and pepper. Remove from heat.
4. Pour mixture over pasta. Add parsley and crab to pasta and toss well.
5. Chill before serving.

Nutritional Analysis
(per serving)

435	Calories
2.6 gm.	Fat
14.9 mg.	Cholesterol
5%	of Calories from Fat
23.6 gm.	Protein
0.35 gm.	Saturated Fat
5.93 gm.	Fiber
78.6 gm.	Carbohydrate
830.6 mg.	Sodium
55 mg.	Calcium

Pot-Roasted Pasta

___
Serves 8
___

Be sure to undercook the pasta because you will be baking it. Baking marries the flavors for a delicious result!

1 tablespoon olive oil
1 medium onion, sliced
2 garlic cloves, minced
1 teaspoon Italian Blend*
1 can (28 ounces) tomato sauce
1 can (28 ounces) whole tomatoes with basil, chopped
3 carrots, julienned
4 stalks celery, sliced
1½ pounds mushrooms, washed and cut into quarters
2¼ ounces ripe olives, sliced
1 tablespoon chopped fresh parsley
1 pound pasta, undercooked

1. Heat olive oil in Dutch oven, sauté onion and garlic until soft. Add Italian Blend and sauté 30 seconds.
2. Add tomato sauce and whole tomatoes. Simmer uncovered for 30 minutes.
3. Add carrots, celery, mushrooms, and olives, and simmer 1 hour.
4. Preheat oven to 350°F.
5. Mix pasta into sauce (put into a baking dish or leave in the Dutch oven), cover, and bake for 30 minutes.

*See the herb blends on page 40.

Nutritional Analysis
(per serving)

343	Calories
7.5 gm.	Fat
0.0 mg.	Cholesterol
20%	of Calories from Fat
11.8 gm.	Protein
1.05 gm.	Saturated Fat
13.27 gm.	Fiber
60.3 gm.	Carbohydrate
520 mg.	Sodium
127 mg.	Calcium

Prosciutto and Tomato Pasta

Serves 6

I keep prosciutto in my freezer, so if I need to make a last-minute meal, I'm ready!

1 tablespoon garlic-infused olive oil
3 ounces prosciutto, diced
2 cans (28 ounces) tomatoes, diced
2 tablespoons fresh basil, chopped (1 teaspoon dried)
1 pound thick pasta, cooked al dente
3 ounces Parmesan cheese, freshly grated

1. In a Dutch oven heat olive oil, and add garlic and prosciutto. Sauté until golden.
2. Add tomatoes and cover, leaving room for the steam to escape so the sauce will thicken. Let simmer for 45 minutes. Add fresh basil.
3. Pour sauce over hot pasta and toss with Parmesan cheese.

Nutritional Analysis (per serving)

425	Calories
9.0 gm.	Fat
18.2 mg.	Cholesterol
19%	of Calories from Fat
19.9 gm.	Protein
3.47 gm.	Saturated Fat
5.92 gm.	Fiber
65.6 gm.	Carbohydrate
444.9 mg.	Sodium
256 mg.	Calcium

Shrimp Scampi Provincial with Linguine

Serves 6

A twist on scampi, with added color and less garlic.

1 pound shrimp (30 to a pound), cleaned and peeled
2 teaspoons olive oil
1 clove garlic, minced
2 shallots, finely chopped
½ teaspoons Italian Blend*
2 celery stalks, finely chopped
1 green bell pepper, finely chopped
1 red bell pepper, finely chopped
1 small carrot, finely chopped
1 pound linguine
1 tablespoon butter
Chopped fresh parsley

1. Rinse shrimp and pat dry.
2. In nonstick skillet, heat olive oil. Add garlic and shallots and cook briefly.
3. Add Italian Blend, celery, peppers, and carrots. Sauté until soft.
4. Add shrimp and cook until just done.
5. Cook linguine until al dente. Toss with butter.
6. Add shrimp mixture into pasta and toss, top with fresh parsley.

*See the herb blends on page 40.

Nutritional Analysis (per serving)

384 Calories
6.0 gm. Fat
65.5 mg. Cholesterol
14% of Calories from Fat
18.4 gm. Protein
1.86 gm. Saturated Fat
4.35 gm. Fiber
62.9 gm. Carbohydrate
88.8 mg. Sodium
50 mg. Calcium

Sun-Dried Tomatoes, Rosemary, and Thyme Pasta

Serves 6

This pasta dish is light, so it's perfect for a warm spring or summer night. It's also a good dish for a buffet; using the shorter pasta makes it easier to eat!

1 tablespoon garlic-infused olive oil

2 shallots, chopped

1 cup dry vermouth

6 small leeks, trimmed, cleaned, and cut into 3/4-inch slices

4 ounces sun-dried tomatoes, steeped in 1/2 cup hot water

2 teaspoons fresh rosemary (1/2 teaspoon dried)

2 teaspoons fresh thyme (1/2 teaspoon dried)

1/2 teaspoon salt

2 tablespoons fresh lemon juice

1 pound gemelli (or other short, tubular pasta), cooked al dente

2 ounces Parmesan cheese, freshly grated

1. Heat skillet with olive oil and sauté shallots until soft.
2. Add vermouth to the skillet and bring to a simmer. Add leeks, tomatoes with water, rosemary, thyme, salt, and lemon juice.
3. Place hot pasta in a large bowl and toss with cheese and sauce. Serve immediately.

Nutritional Analysis (per serving)

449	Calories
6.6 gm.	Fat
7.3 mg.	Cholesterol
13%	of Calories from Fat
16.1 gm.	Protein
2.3 gm.	Saturated Fat
4.45 gm.	Fiber
70.8 gm.	Carbohydrate
347 mg.	Sodium
169 mg.	Calcium

Tomato-Basil Sauce with Mostaccioli

Serves 6

This is not only quick and easy, but it is great to have cold for lunch the next day.

2 cans (28 ounces) whole plum tomatoes with basil
1 tablespoon garlic-infused olive oil
1 onion, diced
½ teaspoon freshly ground black pepper
½ teaspoon ground red pepper
1 pound mostaccioli, cooked al dente
3 tablespoons fresh basil leaves, sliced
½ cup chopped fresh parsley
2 ounces Parmesan cheese, freshly grated

1. Place tomatoes in a food processor or blender. Process until smooth.
2. Heat a large skillet or Dutch oven, add olive oil. Add onions to oil and sauté until soft.
3. Add tomatoes and pepper. Simmer 10 minutes uncovered.
4. Add mostaccioli to sauce and stir until heated through.
5. Toss in fresh basil, parsley, and Parmesan cheese.

Nutritional Analysis
(per serving)

426	Calories
7.4 gm.	Fat
7.3 mg.	Cholesterol
16%	of Calories from Fat
17.1 gm.	Protein
2.39 gm.	Saturated Fat
7.95 gm.	Fiber
74.3 gm.	Carbohydrate
639.3 mg.	Sodium
272 mg.	Calcium

Vermicelli with Basil and Pine Nuts

Serves 6

If you want to make this as a vegetarian entrée, use vegetable broth instead of chicken broth.

2 cloves garlic, minced
1 cup chicken broth
½ cup fresh basil leaves, thinly sliced
1 tablespoon cornstarch mixed in ¼ cup chicken broth
½ cup pine nuts
2 ounces romano or Parmesan cheese, freshly grated
Fresh ground black pepper to taste
1 pound vermicelli, cooked al dente

1. In a large Dutch oven, sprayed with olive oil, lightly sauté the garlic.
2. Add chicken broth and bring to a boil. Reduce heat to simmer, add basil, and stir in cornstarch mixture.
3. In a small skillet, toast pine nuts over medium heat.
4. Add pine nuts, cheese, pepper, and vermicelli to broth, and toss to coat well.

Nutritional Analysis (per serving)

426	Calories
11.6 gm.	Fat
7.3 mg.	Cholesterol
24%	of Calories from Fat
18.1 gm.	Protein
3.14 gm.	Saturated Fat
4.24 gm.	Fiber
64.3 gm.	Carbohydrate
307.1 mg.	Sodium
157 mg.	Calcium

Vermicelli with Cream Cheese and Herbs

___
Serves 6
___

You can have this on the table in the time it takes to cook the pasta!

1 pound vermicelli
1 package (8 ounces) light cream cheese
1 teaspoon Italian Blend*
1 clove garlic, chopped
½ teaspoon freshly ground pepper
⅛ teaspoon cayenne pepper
¾ cup boiling water
2 ounces Parmesan cheese, freshly grated
Chopped fresh parsley to garnish

1. In a large pasta pot, cook vermicelli al dente.
2. While vermicelli is cooking, put all other ingredients into food processor or blender. Blend until smooth.
3. Drain pasta well, and then toss in sauce.
4. Garnish with chopped fresh parsley.

Nutritional Analysis (per serving)

467	Calories
17.2 gm.	Fat
48.3 mg.	Cholesterol
33%	of Calories from Fat
16.6 gm.	Protein
10.19 gm.	Saturated Fat
3.42 gm.	Fiber
60.5 gm.	Carbohydrate
286.5 mg.	Sodium
177 mg.	Calcium

*See the herb blends on page 40.

Seafood

From the Sea

Baked Halibut
Baked Sole with Shrimp and Crabmeat
BBQ Swordfish
Ginger-Orange Halibut
Greek-Style Shrimp
Halibut Cacciatore
Poached Halibut and Peppers
Scallops in White Wine
Shrimp-Stuffed Artichokes
Sole Veronique
Spicy Shrimp

From the Sea

Seafood has less fat and fewer calories than any other animal protein and can also be very versatile. Being brought up on the ocean, I had the advantage of buying fish right off the fishing boats. Now, since I no longer live on the coast, I prefer to buy fish that has been flash frozen. With modern technology, fish can be processed on the ship right after being caught and then flash frozen. This fish can be much more appealing than fish that has been around four or five days. When buying fresh fish, ask to smell it first; if it has a fishy smell or an ammonia scent, don't buy it.

Now I don't want to give you the impression that you have to live on the ocean to get fresh fish. Lake fish is delicious, and there are many fish farms around. Check your area for a store that sells farm-raised fish.

Firm fish, such as halibut and swordfish, are excellent barbecued. Shrimp can be used in many ways; find a frozen brand you like and keep a bag in the freezer.

Baked Halibut

Serves 4

If you want this a bit fancier, make a heavy white sauce and spoon it over the halibut halfway through cooking. Finish the same way with the mushrooms and parsley.

4 halibut steaks, 6 ounces each

1 lemon

1 teaspoon Basic Herb Blend*

1 pound mushrooms, sliced

½ cup dry vermouth

Fresh chopped parsley

1. Preheat oven to 350°F.
2. Trim skin off halibut, rinse, and pat dry. Squeeze lemon on both sides of halibut.
3. Place halibut in a baking dish and sprinkle with Basic Herb Blend. Bake for 20 minutes (or until done).
4. Place mushrooms in a nonstick skillet. Add vermouth, cover, and cook on medium heat for 5 minutes. Uncover and cook until all liquid is absorbed.
5. To serve, place mushrooms on top of the halibut and sprinkle with parsley.

*See the herb blends on page 40.

Nutritional Analysis (per serving)

240	Calories
7.5 gm.	Fat
94.2 mg.	Cholesterol
28%	of Calories from Fat
37.9 gm.	Protein
1.53 gm.	Saturated Fat
0.77 gm.	Fiber
3.6 gm.	Carbohydrate
220.5 mg.	Sodium
46 mg.	Calcium

Baked Sole with Shrimp and Crabmeat

Serves 4

The fishmonger I went to in the 1980s gave me this recipe. I've been making it ever since. You can easily make this for as many people as you want. The flavors of the sole, shrimp, crab, and dill are memorable.

4 large shrimp, uncooked
4 sole fillets, 3 ounces each
4 ounces crabmeat
1 lemon
Paprika
Fresh dill sprigs (or dried dillweed)

1. Preheat oven to 350°F.
2. Shell, clean, and butterfly shrimp.
3. Place a shrimp in the center of each sole fillet. Then, put 1 ounce of crabmeat in each shrimp, roll, and secure with a toothpick.
4. Place fish rolls, seam side down, in a baking dish that has been sprayed with a nonstick spray.
5. Cut 6 thin slices from the center of the lemon, remove seeds, and cut in half. From the end of the lemon, squeeze juice over the fish.
6. Sprinkle with paprika. Lay a dill sprig on top along with half lemon slices.
7. Bake for about 20 minutes.

Nutritional Analysis (per serving)

140	Calories
4 gm.	Fat
74.1 mg.	Cholesterol
26%	of Calories from Fat
23.7 gm.	Protein
0.82 gm.	Saturated Fat
0.21 gm.	Fiber
1.3 gm.	Carbohydrate
175.5 mg.	Sodium
43 mg.	Calcium

BBQ Swordfish

Serves 4

When entertaining, make this along with grilled vegetables, chicken, and steak—and you'll have a great mixed grill with something for everyone.

4 swordfish fillets, 6 ounces each
½ cup low-sodium soy sauce
Juice from 1 lemon
1 tablespoon grated ginger (1 teaspoon dried)
1 clove garlic, chopped
1 teaspoon dry mustard
¼ teaspoon cayenne pepper

1. Trim skin off swordfish, rinse, and pat dry.
2. Mix all other ingredients together in a dish or pan large enough to accommodate the swordfish.
3. Place swordfish in the marinade. Turn to cover both sides. Marinate 2 to 6 hours in the refrigerator, turning once.
4. Cook on a barbecue grill or broil until done.

Nutritional Analysis (per serving)

222	Calories
9.6 gm.	Fat
52.4 mg.	Cholesterol
39%	of Calories from Fat
28.3 gm.	Protein
2.28 gm.	Saturated Fat
0.92 gm.	Fiber
3.6 gm.	Carbohydrate
1541.4 mg.	Sodium
19 mg.	Calcium

Ginger-Orange Halibut

Serves 4

If you have a grilling machine, use it. Remember, it will take only about 4 minutes to cook!

4 halibut steaks, about 5 ounces each
1 lemon, juiced
Freshly ground black pepper, to taste
1 teaspoon ginger, freshly grated
1 tablespoon cornstarch
1 cup chicken broth
1 tablespoon orange rind
1 tablespoon soy sauce
1 green onion, finely chopped

1. Preheat broiler in oven.
2. Place halibut on broiler pan sprayed with nonstick spray, and sprinkle with half of the lemon juice. Grind fresh pepper on top. Broil 4 inches from heat for 4 to 5 minutes.
3. Turn fish over and sprinkle remaining lemon juice over the top. Grind fresh pepper on top and broil another 4 to 5 minutes.
4. While halibut is cooking, mix the remaining ingredients in a saucepan and heat until mixture thickens.
5. Pour sauce over halibut and serve.

Nutritional Analysis
(per serving)

209	Calories
6.3 gm.	Fat
78.5 mg.	Cholesterol
27%	of Calories from Fat
32.1 gm.	Protein
1.33 gm.	Saturated Fat
0.79 gm.	Fiber
4.5 gm.	Carbohydrate
626.1 mg.	Sodium
46 mg.	Calcium

Greek-Style Shrimp

_____

Serves 6

_____

As a first course serve the Phyllo Goat-Cheese Squares (page 57), except use the Feta Cheese Spread (page 45) instead of the goat cheese. This gives continuity to the flavors.

1½ pounds raw shrimp (16 to a pound)
1 tablespoon olive oil
1 medium onion, chopped
½ cup dry vermouth
1 can (27 ounces) diced tomatoes with juice
2 tablespoons chopped fresh parsley
½ teaspoon oregano
¼ teaspoon freshly ground black pepper
¼ pound feta cheese, crumbled

1. Clean and rinse shrimp, pat dry, set aside.
2. Heat oil in large skillet and sauté onion until soft.
3. Stir in vermouth, tomatoes, parsley, oregano, and pepper. Bring to boil and cook over medium heat until mixture thickens slightly.
4. Add shrimp and cook until they are done, about 3 to 5 minutes.
5. Stir in feta cheese and serve.

Nutritional Analysis (per serving)

165   Calories
7.3 gm. Fat
64.6 mg. Cholesterol
40%   of Calories from Fat
10.7 gm. Protein
3.28 gm. Saturated Fat
2.31 gm. Fiber
10.1 gm. Carbohydrate
480 mg. Sodium
157 mg. Calcium

Halibut Cacciatore

Serves 6

Halibut works very well in this because it is so firm. It can be served over pasta or rice.

1 tablespoon olive oil
1 medium onion, sliced
2 cloves garlic, minced
1 teaspoon oregano
1 pound mushrooms, sliced
1 green bell pepper, sliced
2 diced tomatoes
1 can (28 ounces) tomato sauce
½ teaspoon black pepper
2 tablespoons chopped fresh parsley
1 pound halibut, cut into 12 pieces

1. Heat olive oil in Dutch oven. Add onion and garlic, sauté until soft. Add oregano and sauté 30 seconds.
2. Add mushrooms and bell pepper. Cover and simmer 3 minutes. Uncover and cook until liquid is reduced.
3. Add tomatoes, tomato sauce, pepper, and parsley. Bring to simmer. Simmer for 30 minutes.
4. Add halibut, cover, and simmer 10 to 15 minutes.

Nutritional Analysis
(per serving)

186	Calories
6 gm.	Fat
41.9 mg.	Cholesterol
29%	of Calories from Fat
19.4 gm.	Protein
1.05 gm.	Saturated Fat
3.23 gm.	Fiber
15.5 gm.	Carbohydrate
961.9 mg.	Sodium
54 mg.	Calcium

Poached Halibut and Peppers

―――――――
Serves 4
―――――――

This is colorful and tasty. If you can handle the extra calories, finish the peppers with 1 tablespoon of butter.

1 cup dry vermouth
1 bay leaf
4 halibut steaks, 6 ounces each, with skin removed
1 red bell pepper, thinly sliced
1 green bell pepper, thinly sliced
1 yellow bell pepper, thinly sliced
1 medium onion, thinly sliced
2 celery stalks, thinly sliced
Paprika, cayenne pepper, and black pepper to taste

1. In a large skillet, bring vermouth with bay leaf to a boil. Add the halibut steaks and simmer until fish is done. Remove halibut and set aside. Discard bay leaf.

2. Bring liquid to a boil and reduce to $\frac{1}{2}$ cup. Add remaining ingredients and simmer until all liquid is absorbed.

3. Serve vegetable mixture over halibut.

Nutritional Analysis
(per serving)

334	Calories
7.5 gm.	Fat
94.2 mg.	Cholesterol
20%	of Calories from Fat
38.3 gm.	Protein
1.54 gm.	Saturated Fat
1.8 gm.	Fiber
12.3 gm.	Carbohydrate
243.6 mg.	Sodium
65 mg.	Calcium

Scallops in White Wine

Serves 6

Depending on what is available, I prefer the large sea scallops because they are firmer. However, this is a matter of individual taste. Some people prefer the small bay scallops because they are very tender. This dish can be served in a sea shell or on top of buttered pasta.

1½ pounds scallops
2 cups dry white wine or dry vermouth
4 shallots, chopped
2 pounds mushrooms, quartered
2 tablespoons chopped fresh parsley
½ teaspoon salt
½ teaspoon marjoram

1. Wash scallops and pat dry. In large nonstick skillet, simmer scallops in wine for 5 minutes. Remove scallops and set aside.
2. Bring liquid to a boil. Add shallots and mushrooms. Cook until most of the liquid is absorbed.
3. Add scallops, parsley, salt, and marjoram. Serve immediately.

Nutritional Analysis (per serving)

191	Calories
2.6 gm.	Fat
24 mg.	Cholesterol
12%	of Calories from Fat
13.4 gm.	Protein
0.45 gm.	Saturated Fat
0.66 gm.	Fiber
8.7 gm.	Carbohydrate
311 mg.	Sodium
34 mg.	Calcium

Shrimp-Stuffed Artichokes

Serves 4

This dish presents exceptionally well. It is also fun to eat and lends itself well to conversation. It is a good luncheon dish or a first course for a dinner party.

4 artichokes, cooked
8 ounces cooked shrimp, cubed
2 celery stalks, diced
½ cup light sour cream
¼ cup low-calorie mayonnaise
1 tablespoon dillweed
⅛ teaspoon garlic powder
4 large red lettuce leaves
1 can (4 ounces) mandarin oranges

1. Remove the center core of the artichokes and scrape the remaining center leaves clean. Trim the points off the leaves and chill.
2. Mix shrimp, celery, sour cream, mayonnaise, dillweed, and garlic powder. Chill for 1 hour.
3. Fill artichokes with shrimp mixture and place on individual serving plates lined with red lettuce. Garnish with mandarin oranges, putting some oranges between the leaves of the artichokes.

Nutritional Analysis
(per serving)

164	Calories
6.6 gm.	Fat
34 mg.	Cholesterol
36%	of Calories from Fat
8.6 gm.	Protein
2.48 gm.	Saturated Fat
7.17 gm.	Fiber
21.6 gm.	Carbohydrate
246.8 mg.	Sodium
107 mg.	Calcium

Sole Veronique

Serves 6

This is a version of the classic French dish, modified for ease and lower fat content.

1½ pounds sole fillets
1 cup dry white wine or dry vermouth
1 teaspoon chervil
⅛ teaspoon cayenne pepper (or to taste)
1 tablespoon cornstarch
1 cup whole milk
1 cup seedless green grapes, cut in half (about 30 grapes)
Paprika to garnish

1. Preheat oven to 350°F.
2. Rinse sole and pat dry.
3. In a skillet bring wine, chervil, and cayenne pepper to a simmer. Gently add sole, simmer for 2 minutes on each side. Transfer fish to a baking/serving dish.
4. Mix cornstarch into milk, pour into skillet, and heat, stirring constantly until thickened. Add grapes to sauce.
5. Pour sauce over sole and bake at 350°F until flaky (about 10 minutes).
6. Sprinkle with paprika and serve.

Nutritional Analysis (per serving)

252	Calories
6.4 gm.	Fat
68.3 mg.	Cholesterol
23%	of Calories from Fat
26.3 gm.	Protein
1.9 gm.	Saturated Fat
0.21 gm.	Fiber
11.5 gm.	Carbohydrate
170 mg.	Sodium
82 mg.	Calcium

Spicy Shrimp

Serves 4

Perfect with plain rice and snow peas. You can double this recipe if company is coming or use it as a first course for six to eight people by serving it in lettuce cups.

1 teaspoon olive oil
1 pound raw shrimp, cleaned and peeled
2 green onions, chopped
1 clove garlic, minced
2 tablespoons fresh ginger, minced
2 tablespoons sherry
2 tablespoons soy sauce
¼ cup ketchup
½ teaspoon crushed red pepper (or to taste)

1. Heat oil in a skillet, add shrimp, onions, garlic, and ginger. Quickly stir-fry until shrimp is pink.
2. Add remaining ingredients and stir well.

Nutritional Analysis (per serving)

132	Calories
2 gm.	Fat
54 mg.	Cholesterol
13%	of Calories from Fat
8.6 gm.	Protein
0.35 gm.	Saturated Fat
2.75 gm.	Fiber
10.6 gm.	Carbohydrate
767.1 mg.	Sodium
41 mg.	Calcium

Poultry

A Chicken in Every Pot

Marinade
Soy-Ginger Marinade
Baked Chicken
BBQ Chicken with Vinegar-Basil Marinade
Cajun Chicken
Chicken Breast Supreme
Chicken Cacciatore
Chicken Cakes with Tomato and Sweet Pepper Sauce
Chicken Chasseur
Chicken in Lettuce Cups
Chicken Polynesian
Chicken Polynesian II
Chicken Stew
Chicken Strips with Brussels Sprouts
Chicken Thighs with an Indian Flair
Chicken with Lemon and Capers
Mustard Chicken
Orange and Onion Chicken
Orange-Rosemary Cornish Game Hens
Oven-Fried Chicken
Rolled Chicken with Asparagus
Turkey Breast Marinated with Lemon and Herbs

Turkey Chili
Turkey Sausage
Country Sausage Gravy
Turkey Sausage with Pepper and Onions

A Chicken in Every Pot

There was a time in US history when Sunday dinner was always chicken. Chicken has now become the most frequently served animal protein in most homes. One reason for its popularity is that chicken is so versatile. With chicken in your freezer, you will certainly have a great meal ahead. I do a lot of recipes with boneless chicken breast because they are quick, easy, and healthful. However, I do prefer the flavor of cooking the breast with the bone in and the skin off! You can also find fresh or frozen chicken thighs that are already boned and skinned. I prefer using the thigh meat in stews or recipes that require a longer cooking time; the meat stays moist and has a great flavor. Poultry can take on many flavors, depending on the herbs and spices—from a spicy Cajun dish to a delicately seasoned French entrée.

To Marinade—To Top

The two basic marinades given here can be used with most poultry and meat items. If you need to marinate quickly, put the marinade in an airtight, lock-top plastic bag. With a fork, poke some holes in the meat and put it into the bag. Try to get as much air out of the bag as you can, then place it in the refrigerator. Turn the bag over every 5 to 10 minutes. This will do a pretty good job of infusing the flavors in 15 to 30 minutes.

For a topping, keeping Yogurt Cheese (page 227) around can be a saving grace, and the creation of Yogurt Cream (page 228) is one of the simplest and easiest I have come up with. Yogurt Cream is a perfect topping for desserts because it has no fat and tastes wonderful. You can also add extracts to it to enhance the flavor of what you're topping. For example, try adding lemon extract to top a lemon cake.

Marinade

Try marinating chicken pieces or a leg of lamb in this delicious sauce. Have the butcher bone and butterfly the leg. It's best to cook it on an outdoor grill if you have one.

Juice of 1 lemon
½ teaspoon minced garlic
1 teaspoon olive oil
⅛ teaspoon coriander
⅛ teaspoon cumin
⅛ teaspoon cayenne pepper

Combine all ingredients.

Nutritional Analysis (per serving)

53 Calories
4.5 gm. Fat
0.0 mg. Cholesterol
76% of Calories from Fat
0.6 gm. Protein
0.58 gm. Saturated Fat
1.02 gm. Fiber
4.6 gm. Carbohydrate
1.6 mg. Sodium
17 mg. Calcium

Soy-Ginger Marinade

Will marinate 1 pound of chicken or beef.

This is my favorite all-around marinade for chicken and beef. Put the marinade and meat in a lock-top plastic bag and refrigerate it for ½ to 2 hours, turning it occasionally.

Juice of 1 lemon

2 tablespoons soy sauce

2 tablespoons water

½ teaspoon minced garlic

¼ teaspoon ground ginger

½ teaspoon paprika

Mix all ingredients together.

Nutritional Analysis (per serving)

95	Calories
0.4 gm.	Fat
0.0 mg.	Cholesterol
4%	of Calories from Fat
8.2 gm.	Protein
0.03 gm.	Saturated Fat
1.44 gm.	Fiber
17 gm.	Carbohydrate
8231.2 mg.	Sodium
40 mg.	Calcium

Baked Chicken

Serves 6

There are great ready-made polentas you can buy. Try them with this dish—just slice the polenta and grill them.

2 teaspoons rosemary
½ teaspoon Italian Blend*
3 cloves garlic, crushed
6 chicken breasts, boneless and skinless
½ cup dry vermouth
1 cup low-sodium chicken broth
Fresh parsley for garnish

1. Preheat oven to 400°F.
2. Mix rosemary, Italian Blend, and garlic together well in a mortar and pestle or minichopper.
3. Rub the seasonings on the chicken breasts, place in a baking dish, and bake for 15 minutes.
4. Reduce the oven temperature to 350°F. Pour the vermouth and chicken broth over the chicken. Cover and bake 20 to 30 minutes longer (until done).
5. Place chicken on a serving platter and garnish with fresh parsley.

Nutritional Analysis
(per serving)

172	Calories
3.2 gm.	Fat
72.7 mg.	Cholesterol
17%	of Calories from Fat
27.1 gm.	Protein
0.91 gm.	Saturated Fat
0.1 gm.	Fiber
1.9 gm.	Carbohydrate
196.2 mg.	Sodium
25 mg.	Calcium

*See the herb blends on page 40.

BBQ Chicken with Vinegar-Basil Marinade

Serves 4

Partially cooking the chicken in marinade helps speed up the cooking time on the grill and keeps the meat moist. If you want to use the marinade for basting, be sure to bring it to a boil before reusing it in order to kill any bacteria.

4 chicken thighs with legs, skinned
¾ cup low-salt chicken broth
2 tablespoons herb vinegar
½ cup dry vermouth
2 shallots, chopped
¼ teaspoon freshly ground black pepper
2 teaspoons dried basil

1. Rinse and pat chicken dry.
2. Bring remaining ingredients to a simmer in a skillet large enough to hold the chicken.
3. Add the chicken to the sauce, cover, and simmer 20 minutes, turning once.
4. Remove from heat and refrigerate for 4 to 8 hours (or overnight).
5. Cook chicken on a barbecue until done, or in the broiler, about 8 minutes on each side.

Nutritional Analysis (per serving)

193	Calories
3.4 gm.	Fat
77.2 mg.	Cholesterol
16%	of Calories from Fat
28.8 gm.	Protein
0.96 gm.	Saturated Fat
0.11 gm.	Fiber
2.5 gm.	Carbohydrate
217.3 mg.	Sodium
31 mg.	Calcium

Cajun Chicken

Serves 4

Cajun recipes have French and Spanish roots and are wonderfully spicy. Try the same preparation with pork, flank steak, shrimp, or halibut.

8 chicken thighs, skin and fat removed
½ teaspoon caraway seed
½ teaspoon cayenne pepper (or to taste)
¾ teaspoon ground coriander
¾ teaspoon ground cumin
3 cloves garlic, minced
1½ teaspoons dry mustard
¼ teaspoon thyme
¼ cup cognac
2 tablespoons fresh lemon juice

1. Preheat oven to 350°F.
2. Rinse and pat chicken dry.
3. Place all ingredients, except the chicken, into a minichopper and pulverize.
4. Line a baking sheet with foil. Coat each piece of chicken with the spice mixture, and place on the baking sheet. Bake for 35 to 45 minutes.

Nutritional Analysis
(per serving)

339	Calories
7 gm.	Fat
149.9 mg.	Cholesterol
18%	of Calories from Fat
55.4 gm.	Protein
1.81 gm.	Saturated Fat
0.38 gm.	Fiber
2.3 gm.	Carbohydrate
132.8 mg.	Sodium
46 mg.	Calcium

Chicken Breast Supreme

Serves 4

This dish is excellent served with rice pilaf and green beans. Try topping it with sliced black olives; they add a nice flavor and color.

1 pound large mushrooms, sliced
2 tablespoons butter
4 chicken breasts, boned and skin removed (4 to 5 ounces each)
¼ cup unbleached flour
½ teaspoon salt
½ teaspoon white pepper
½ cup dry white vermouth
1½ cups low-sodium chicken broth
¼ cup cold water with 2 teaspoons cornstarch mixed in
½ cup light sour cream
¼ cup chopped fresh parsley

1. Sauté mushrooms in 1 tablespoon butter, set aside.
2. Rinse chicken breasts in water and pat dry.
3. Put the flour, salt, and pepper into a bag. Toss the chicken in the mixture until well coated.
4. Heat the remaining butter in a skillet, and brown the chicken on both sides.
5. Add vermouth to deglaze the pan, cook 2 minutes.
6. Add the chicken broth and bring to a boil. Reduce to simmer, cover, and simmer for 20 minutes. Remove chicken and keep warm.
7. Bring liquid to a boil, and add the cornstarch-water mixture to thicken.
8. Add the sour cream and mix in, but do not boil. Add the chicken back into the sour cream mixture and heat through.
9. To serve, place the chicken on a plate and spoon the sauce over it. Top with mushrooms and garnish with parsley.

Nutritional Analysis (per serving)

332	Calories
13 gm.	Fat
95.4 mg.	Cholesterol
35%	of Calories from Fat
31.5 gm.	Protein
6.69 gm.	Saturated Fat
1.81 gm.	Fiber
14.9 gm.	Carbohydrate
682.2 mg.	Sodium
68 mg.	Calcium

Chicken Cacciatore

Serves 6

I've found that dark meat works better in this recipe because it stays moist. But feel free to add some breast to it, bone in, skin off!

1 tablespoon olive oil
6 chicken legs with thighs, skin removed (about 4 ounces each)
1 medium onion, sliced
2 cloves garlic, minced
1 teaspoon oregano
1 pound mushrooms, sliced
1 green bell pepper, sliced
1 can (14½ ounces) diced tomatoes
1 can (28 ounces) tomato sauce
½ teaspoon black pepper
2 tablespoons fresh parsley, chopped

1. Heat olive oil in Dutch oven, brown chicken legs, and remove from pan.
2. Add onion and garlic, sauté until soft. Add oregano and sauté 30 seconds.
3. Add mushrooms and bell pepper. Cover and simmer 3 minutes. Uncover and cook until liquid is reduced.
4. Add tomatoes, tomato sauce, pepper, and parsley. Bring to simmer.
5. Add chicken legs, cover, and simmer 45 minutes.

Nutritional Analysis (per serving)

246	Calories
6.2 gm.	Fat
75.7 mg.	Cholesterol
23%	of Calories from Fat
31.7 gm.	Protein
1.3 gm.	Saturated Fat
3.91 gm.	Fiber
17.7 gm.	Carbohydrate
87.7 mg.	Sodium
53 mg.	Calcium

Chicken Cakes with Tomato and Sweet Pepper Sauce

Serves 8

This recipe takes longer to prepare, but it is so good! This dish is great for a buffet. To have as a first course, make the cakes smaller.

TOMATO AND SWEET PEPPER SAUCE

2 shallots, chopped

1 red bell pepper, seeded and chopped

3 cans (14½ ounces) diced tomatoes in juice

¼ cup dry white vermouth

1 tablespoon fresh tarragon (or 1 teaspoon dried)

½ cup chopped fresh parsley

½ teaspoon cayenne pepper

1. Spray a skillet with a nonstick spray, add shallots and red bell peppers and sauté until tender.
2. Add tomatoes and simmer, uncovered, 8 to 10 minutes, stirring occasionally.
3. Stir in wine and continue simmering until sauce thickens (about 15 minutes).
4. Add tarragon, parsley, and cayenne pepper. Set aside while you form the chicken cakes.

(continued on next page)

Nutritional Analysis
(per serving)

290	Calories
4.6 gm.	Fat
73 mg.	Cholesterol
14%	of Calories from Fat
33.2 gm.	Protein
1.12 gm.	Saturated Fat
4.16 gm.	Fiber
27.1 gm.	Carbohydrate
515.9 mg.	Sodium
166 mg.	Calcium

CHICKEN CAKES

8 chicken breasts, boned and skinned
½ cup chopped celery
6 green onions, chopped
1 cup chopped red pepper
2 cups dried bread crumbs, finely chopped
2 egg whites
½ cup nonfat milk
2 tablespoons chopped fresh parsley
2 tablespoons chopped fresh tarragon (or 2 teaspoons dried)
½ teaspoon ground black pepper

1. Put chicken breast into a food processor. With a quick on/off, coarsely chop the meat.
2. In a skillet sprayed with nonstick spray, sauté celery, green onion, and red pepper until tender.
3. In mixing bowl, add chicken, vegetables, bread crumbs, egg whites, milk, parsley, tarragon, and pepper. Mix together well and chill for at least 2 hours.
4. Preheat oven to 400°F.
5. Form chicken mixture into 16 cakes. Place on cookie sheet sprayed with a nonstick spray. Bake for 15 minutes. Remove cakes, and reduce temperature to 350°F.
6. Place half of the tomato and sweet pepper sauce into a cook-and-serve baking dish. Place the cakes on the sauce and bake, covered, for ½ hour.
7. Either serve from the baking dish at the table or place on individual plates (2 cakes per serving). Pass the remaining sauce at the table.

Chicken Chasseur

Serves 6

Derived from a classic French dish, Chicken Chasseur is good served with Mashed Potatoes (page 191) and a green vegetable.

6 chicken breast halves, boned and skinned
1 pound mushrooms, sliced
½ cup dry vermouth (or white wine)
1 medium onion, sliced
1 can (28 ounces) whole tomatoes (puréed in blender)
2 teaspoons tarragon
2 teaspoons chervil
½ teaspoon ground black pepper
Chopped fresh parsley for garnish.

1. Spray a large skillet with olive-oil spray. Heat skillet and brown chicken breasts on both sides, then remove from pan.
2. Add mushrooms to pan along with ¼ cup vermouth. Cover and simmer for 2 minutes. Remove cover, turn up heat, and cook until all liquid is absorbed. Remove mushrooms from pan and set aside.
3. Add onion to pan and sauté until soft. Add remaining vermouth.
4. Return mushrooms to pan. Add tomatoes, tarragon, chervil, and pepper. Bring to a boil, reduce heat to simmer.
5. Place chicken in the pan, cover, leaving a small space for the steam to escape to reduce the sauce. Simmer for 30 minutes.
6. Garnish with parsley and serve.

Nutritional Analysis (per serving)

223	Calories
3.8 gm.	Fat
72.7 mg.	Cholesterol
15%	of Calories from Fat
29.9 gm.	Protein
0.96 gm.	Saturated Fat
3.35 gm.	Fiber
12.9 gm.	Carbohydrate
297.9 mg.	Sodium
71 mg.	Calcium

Chicken in Lettuce Cups

Serves 6

This dish is great as a finger food, for picnics, and on warm summer nights. It is best to chop the chicken breast yourself. Use the food processor and cut the breast in large chucks, then pulse it a few times.

1 tablespoon cornstarch
3 tablespoons dry sherry
6 chicken breasts, finely chopped
2 cloves garlic, chopped
2 scallions, diced
8 ounces mushrooms, chopped
1 tablespoon water
1 tablespoon freshly chopped ginger
1 tablespoon dark sesame oil
½ cup water chestnuts, chopped
½ teaspoon freshly ground black pepper
⅛ tablespoon cayenne pepper
3 tablespoons low-sodium soy sauce
12 lettuce cups (iceberg or bibb)

Nutritional Analysis
(per serving)

200	Calories
5.5 gm.	Fat
72.7 mg.	Cholesterol
25%	of Calories from Fat
27.7 gm.	Protein
1.22 gm.	Saturated Fat
1.35 gm.	Fiber
5.8 gm.	Carbohydrate
561.6 mg.	Sodium
23 mg.	Calcium

1. Combine cornstarch and sherry and mix into chicken. Cover and put in refrigerator to marinate for 15 minutes.
2. In large wok or skillet sprayed with olive oil, sauté garlic and scallions lightly. Add mushrooms and 1 tablespoon water. Stir and cover for 3 minutes. Remove cover and cook until all water is gone. Add ginger, then remove from pan and set aside.
3. Heat sesame oil in wok or skillet, add chicken, and stir-fry quickly.
4. Add mushroom mixture, water chestnuts, black and cayenne peppers, and soy sauce. Mix well.
5. Divide the mixture into the 12 lettuce cups and serve.

Chicken Polynesian

Serves 4

The fresh pineapple makes this dish outstanding, and the good part is most markets carry fresh pineapple already cut up and in airtight bags.

1 cup chopped fresh pineapple
½ cup sliced green onion
¼ cup chopped celery
1 tablespoon grated fresh ginger
4 chicken breasts, skin removed
1 tablespoon olive oil
2 tablespoons soy sauce

1. Preheat oven to 350°F.
2. Cover the bottom of a baking dish with pineapple, green onions, celery, and ginger. Lay chicken on top.
3. In a small bowl, mix olive oil and soy sauce. Pour over chicken, cover, and bake for 35 to 45 minutes.

Nutritional Analysis (per serving)

224	Calories
6.7 gm.	Fat
72.7 mg.	Cholesterol
27%	of Calories from Fat
27.7 gm.	Protein
1.37 gm.	Saturated Fat
1.92 gm.	Fiber
11.7 gm.	Carbohydrate
567 mg.	Sodium
37 mg.	Calcium

Chicken Polynesian II

Serves 4

This is a variation on Chicken Polynesian. This time you're going to use ingredients that you have on hand in the cupboard and freezer.

4 chicken breasts, boned and skinned
¼ cup low-sodium soy sauce
1 teaspoon ginger powder
1 teaspoon onion powder
½ cup all-purpose flour
1 can (4 ounces) pineapple chunks
1 can (4 ounces) mandarin oranges
2 tablespoons cornstarch
¼ cup water
¼ cup toasted almonds, chopped

1. Rinse and pat chicken dry. Place in a shallow pan.
2. Combine soy sauce, ginger, and onion powder, and pour over chicken. Cover and marinate in refrigerator 1 to 2 hours.
3. Put flour in a bag, add chicken breasts, and shake until they are well coated.
4. Spray a large nonstick skillet with nonstick spray, and brown chicken on both sides. Return chicken to baking pan.
5. Preheat oven to 350°F.
6. Drain pineapple and mandarin oranges, reserving juice. Set aside the fruit. Add the juice to the marinade mixture and pour into the skillet.
7. Combine cornstarch and water and pour into the skillet. Stir over heat until sauce is thickened. Pour sauce over chicken and bake 20 minutes.
8. Add fruit and almonds to chicken dish and bake another 10 minutes.

Nutritional Analysis (per serving)

299	Calories
7.4 gm.	Fat
72.7 mg.	Cholesterol
22%	of Calories from Fat
30.8 gm.	Protein
1.29 gm.	Saturated Fat
2.05 gm.	Fiber
26.4 gm.	Carbohydrate
1152.7 mg.	Sodium
51 mg.	Calcium

Chicken Stew (The Easy Way)

Serves 6

This great-tasting stew can be made in your crock-pot, cooked on low for 8 hours.

These are all ingredients that you can pull from your pantry and freezer. You can substitute fresh carrots, celery, potatoes, and onions for the frozen stew vegetables, which makes it even better. If you have a bread maker, make a loaf of bread and set it to be ready at the same time as the stew. Come home from work to a ready-made dinner; just add a salad!

8 chicken thighs, boned and skinned
2 packages (10 ounces) frozen stew vegetables
½ cup barley
1 can (10½ ounces) condensed tomato soup
2 cups chicken broth
1½ teaspoons Vegetable Blend*
½ teaspoon freshly ground black pepper
⅛ teaspoon cayenne pepper

1. Preheat oven to 300°F.
2. Rinse chicken, remove excess fat, and cut in large chunks.
3. In a casserole, mix all ingredients, cover, and bake for 4 hours.

*See the herb blends on page 40.

Nutritional Analysis (per serving)

273	Calories
4.4 gm.	Fat
75.7 mg.	Cholesterol
14%	of Calories from Fat
33.2 gm.	Protein
1.19 gm.	Saturated Fat
7.46 gm.	Fiber
25.9 gm.	Carbohydrate
711.1 mg.	Sodium
60 mg.	Calcium

Chicken Strips with Brussels Sprouts

Serves 4

If you don't love Brussels sprouts, substitute zucchini sliced on the diagonal. Serve with roasted quartered potatoes.

1 package (10 ounces) frozen Brussels sprouts, thawed
4 chicken breasts, skinned and cut into strips
¼ teaspoon salt
½ freshly ground black pepper
1 tablespoon garlic-infused olive oil
1 small onion, thinly sliced
1½ tablespoons lemon juice
¾ teaspoon crushed basil
2 cups tomatoes, coarsely chopped

1. Put Brussels sprouts in a colander and run hot water over them, drain, and halve. Set aside.
2. Season chicken with salt and pepper.
3. In a skillet, heat oil and cook chicken strips and onion on medium-high heat until done.
4. Stir in Brussels sprouts, lemon juice, and basil. Reduce heat and cook for 10 minutes.
5. Stir in tomatoes. Cook for 2 minutes more.

Nutritional Analysis (per serving)

214	Calories
7.1 gm.	Fat
72.7 mg.	Cholesterol
30%	of Calories from Fat
28.9 gm.	Protein
1.44 gm.	Saturated Fat
3.8 gm.	Fiber
9.3 gm.	Carbohydrate
204.8 mg.	Sodium
43 mg.	Calcium

Chicken Thighs with an Indian Flair

_____

Serves 4

_____

Thighs are excellent to use in many chicken recipes; just skin them and remove any visible fat. Serve this dish with Indian Raisin Rice (page 198).

8 chicken thighs, skinned
1 cup plain nonfat yogurt
2 tablespoons fresh lemon juice
2 teaspoons freshly grated ginger
2 cloves garlic, minced
1 teaspoon ground cumin
1 teaspoon turmeric
½ teaspoon cayenne pepper
1 teaspoon cornstarch, mixed with 2 tablespoons cold water

1. Rinse chicken thighs and pat dry.
2. Mix remaining ingredients, except cornstarch. Place thighs in yogurt mixture and marinate in refrigerator, covered, for 8 hours.
3. Preheat oven to 350°F.
4. Place thighs in a baking pan sprayed with nonstick spray and bake for 15 minutes.
5. Place the marinade in a saucepan and bring to a simmer. Add the cornstarch mixture, stirring until slightly thickened.
6. Pour the marinade over the chicken thighs and bake for 15 minutes more. Then serve.

Nutritional Analysis (per serving)

339	Calories
6.7 gm.	Fat
151 mg.	Cholesterol
18%	of Calories from Fat
58.6 gm.	Protein
1.87 gm.	Saturated Fat
0.91 gm.	Fiber
6.9 gm.	Carbohydrate
179.9 mg.	Sodium
159 mg.	Calcium

Chicken with Lemon and Capers

Serves 6

Serve this with spinach pasta tossed in butter and freshly grated Parmesan cheese.

- 1 tablespoon olive oil
- 1 medium onion, sliced
- 6 chicken breasts, boned and skinned (about 4 ounces each)
- ¼ cup flour with ½ teaspoon white pepper and ⅛ teaspoon salt mixed in
- 2 cups dry vermouth
- 2 lemons, thinly sliced with skins (remove seeds)
- 2 tablespoons small capers
- ¼ cup chopped fresh parsley

1. In a large skillet heat olive oil, add onion and sauté until tender.
2. Coat chicken breasts with the flour mixture and brown on both sides.
3. Add vermouth and bring to a quick boil. Reduce heat to simmer.
4. Add lemons and capers. Cover and simmer until chicken is done, about 8 to 10 minutes. Add parsley.

Nutritional Analysis (per serving)

294	Calories
5.6 gm.	Fat
72.7 mg.	Cholesterol
17%	of Calories from Fat
27.8 gm.	Protein
1.19 gm.	Saturated Fat
0.92 gm.	Fiber
12 gm.	Carbohydrate
231.7 mg.	Sodium
43 mg.	Calcium

Mustard Chicken

Serves 6

For mustard lovers everywhere, this is sure to please. Serve a colorful vegetable or your table will lack eye appeal.

6 chicken breasts, boneless and skinless
½ cup Dijon mustard
¼ cup white wine vinegar
1 clove garlic, crushed
⅓ teaspoon crushed thyme

SAUCE
1 cup low-sodium chicken broth
1 tablespoon cornstarch
½ cup low-fat sour cream
½ cup Dijon mustard

1. Combine ½ cup mustard, wine vinegar, garlic, and thyme. Pour over chicken to marinate 2 to 6 hours.
2. Grill or broil until done. Place on serving platter and keep warm.
3. To make the sauce, mix the cornstarch into the chicken broth. Heat until thickened, stir in sour cream and mustard.
4. Spoon sauce over chicken breasts.

Nutritional Analysis
(per serving)

202	Calories
6.9 gm.	Fat
77 mg.	Cholesterol
31%	of Calories from Fat
29.2 gm.	Protein
2.24 gm.	Saturated Fat
1.11 gm.	Fiber
5 gm.	Carbohydrate
700.2 mg.	Sodium
62 mg.	Calcium

Orange and Onion Chicken

_____

Serves 6

_____

The oranges and the onions give this chicken an interesting change of taste. For a complementary salad, toss mixed greens with seasoned rice wine vinegar and top with sliced oranges and avocados.

6 chicken breast halves with bone, skinned
2 tablespoons flour
½ teaspoon Vegetable Blend*
⅛ teaspoon salt
¼ teaspoon freshly ground black pepper
½ cup dry white wine or dry vermouth
2 onions, thinly sliced
1 teaspoon orange zest
1 cup orange juice
2 teaspoons fresh thyme (½ teaspoon dried)

1. Rinse chicken breasts and pat dry.
2. Mix flour, Vegetable Blend, and salt, put into a bag. Put chicken breast into bag and shake to coat.
3. In a skillet sprayed with olive oil, brown the chicken. Remove chicken from the skillet and place into a baking pan that has been sprayed with a nonstick spray.
4. Deglaze skillet with ¼ cup of the wine. Add onions and cook until soft.
5. Add orange zest, orange juice, thyme, and remaining wine and bring to a boil, reduce to half. While the sauce is cooking, preheat oven to 350°F.
6. Once sauce is reduced, pour over chicken and bake uncovered for 45 minutes or until done.

_____

*See the herb blends on page 40.

Nutritional Analysis (per serving)

200	Calories
3.2 gm.	Fat
72.7 mg.	Cholesterol
14%	of Calories from Fat
27.3 gm.	Protein
0.89 gm.	Saturated Fat
0.78 gm.	Fiber
9.2 gm.	Carbohydrate
106 mg.	Sodium
27 mg.	Calcium

Orange-Rosemary Cornish Game Hens

Serves 4

Did you know Cornish game hens are all white meat? They are so easy to make and always seem festive. As a side dish, serve Wild-Brown Rice (page 200).

2 Cornish game hens, rinsed inside and out, patted dry
2 cloves garlic, peeled
1 orange, quartered
4 springs fresh rosemary (1 tablespoon dry)
2 tablespoons sugar-free orange marmalade
1 tablespoon chopped fresh rosemary (1½ teaspoons dry)

1. Preheat oven to 350°F.
2. In each hen, place 1 clove garlic, 2 quarters of the orange, and 2 springs of rosemary.
3. Place hens in roasting pan and roast for 70 minutes.
4. Mix orange marmalade and chopped rosemary and baste hens. Roast 10 minutes longer.
5. Let hens rest 10 minutes, than cut in half and serve.

Nutritional Analysis (per serving)

397	Calories
20.9 gm.	Fat
133.9 mg.	Cholesterol
47%	of Calories from Fat
41.7 gm.	Protein
5.77 gm.	Saturated Fat
0.35 gm.	Fiber
8.7 gm.	Carbohydrate
132.1 mg.	Sodium
46 mg.	Calcium

Oven-Fried Chicken

Serves 4

A great alternative to old-fashioned fried chicken! The secret to getting the crunch is to spray the chicken lightly with olive oil before putting it into the oven.

4 chicken breast halves, boned and skinned
½ teaspoon freshly ground black pepper
Pinch of salt
2 tablespoons Dijon mustard
½ cup light sour cream
2 teaspoons Vegetable Blend*
2 tablespoons chopped fresh parsley
1 cup bread crumbs

1. Rinse and pat chicken dry.
2. Preheat oven to 500°F.
3. Mix pepper, salt, mustard, sour cream, Vegetable Blend, and parsley.
4. Coat the chicken on both sides with the mixture and then dip chicken in bread crumbs.
5. Place chicken on baking sheet lined with foil and spray lightly with olive oil.
6. Bake for 30 minutes.

Nutritional Analysis
(per serving)

255	Calories
7.3 gm.	Fat
79.1 mg.	Cholesterol
26%	of Calories from Fat
29.8 gm.	Protein
2.94 gm.	Saturated Fat
1.18 gm.	Fiber
16.1 gm.	Carbohydrate
313.3 mg.	Sodium
74 mg.	Calcium

*See the herb blends on page 40.

Rolled Chicken with Asparagus

Serves 6

This is an excellent dish to serve to company. The coating keeps the chicken moist. Try spooning an Alfredo sauce on top and garnishing with chopped ripe black olives for color.

¼ cup Dijon mustard
1 small clove garlic, minced
¼ cup dry vermouth
6 chicken breasts, boned, skinned, and pounded
18 medium asparagus spears, bottoms trimmed, blanched
1 cup bread crumbs
1 tablespoon grated Parmesan cheese
1 teaspoon Italian Blend*
2 tablespoons chopped fresh parsley

1. Preheat oven to 350°F.
2. Mix mustard, garlic, and vermouth.
3. Coat both sides of chicken breast with mixture. Roll chicken around 3 asparagus spears, and secure with toothpick.
4. Mix bread crumbs, Parmesan cheese, Italian Blend, and parsley. Roll chicken rolls in the mixture.
5. Place chicken in a baking dish sprayed with olive oil. Lightly spray the tops of the chicken rolls with olive oil and bake for 30 minutes.

*See the herb blends on page 40.

Nutritional Analysis (per serving)

251	Calories
4.4 gm.	Fat
73.4 mg.	Cholesterol
16%	of Calories from Fat
30.2 gm.	Protein
1.18 gm.	Saturated Fat
1.54 gm.	Fiber
13.9 gm.	Carbohydrate
310.1 mg.	Sodium
68 mg.	Calcium

Turkey Breast Marinated with Lemon and Herbs

Serves 8

Although the turkey can be roasted in the oven, it is much better when cooked on an outdoor grill.

2 pounds turkey breast, boneless, skinless
Juice and zest of 2 lemons
1 teaspoon dried oregano
Pinch cumin
1 tablespoon chopped fresh parsley
½ teaspoon thyme
½ teaspoon ground white pepper

1. Rinse and pat turkey dry.
2. Mix remaining ingredients.
3. Coat all sides of the turkey with marinade. Cover and refrigerate overnight if possible or at least 4 hours.
4. Preheat oven to 325°F.
5. Roast covered for 35 minutes per pound. Take the cover off for the last 10 minutes. The temperature inside the turkey should be 170°F.
6. Let stand 10 minutes before slicing.

Nutritional Analysis (per serving)

114	Calories
5.2 gm.	Fat
43.3 mg.	Cholesterol
41%	of Calories from Fat
15 gm.	Protein
1.51 gm.	Saturated Fat
0.27 gm.	Fiber
1.2 gm.	Carbohydrate
37 mg.	Sodium
21 mg.	Calcium

Turkey Chili

Serves 8

The perfect chili for a cold winter night or a Super Bowl party!

1 pound dried beans: black, pinto, navy, or mixed
2 teaspoons beef bouillon paste
1 tablespoon garlic olive oil
2 medium onions, chopped
2 bell peppers, chopped
½ teaspoon oregano
¼ teaspoon cumin
3 teaspoons chili powder
1 teaspoon ground red pepper (or to taste)
1 pound ground turkey
1 can (28 ounces) diced tomatoes
1 can (15 ounces) tomato sauce

1. Soak beans overnight in cold water. Rinse beans, put in a large Dutch oven filled with fresh cold water. Bring to a boil, rinse again, and cover beans with fresh water.
2. Add bouillon paste, bring to a boil, and reduce to simmer. Cook until beans are just tender, adding water when needed. When the beans are done, most of the water should be gone.
3. Heat olive oil in a large skillet. Add onions and bell peppers, sauté until soft.
4. Add oregano, cumin, chili powder, and red pepper. Cook 1 minute.
5. Add ground turkey, stir constantly until done.
6. Add turkey mixture to cooked beans. Then add tomatoes and tomato sauce. Simmer 1 hour.

Nutritional Analysis (per serving)

338	Calories
6 gm.	Fat
24.8 mg.	Cholesterol
16%	of Calories from Fat
24.7 gm.	Protein
1.32 gm.	Saturated Fat
13.29 gm.	Fiber
49.6 gm.	Carbohydrate
518 mg.	Sodium
130 mg.	Calcium

Turkey Sausage

Serves 4

Who says you can't enjoy sausage? This has all the flavor of sausage with very little fat. You can also freeze the patties and use them in other recipes.

1 pound ground turkey (mixture of dark and light meat)
½ teaspoon marjoram
½ teaspoon sage
½ teaspoon crushed red pepper (or to taste)
½ teaspoon ground black pepper
½ teaspoon salt

1. If you are grinding your own turkey, cut it into chunks and put it and all the other ingredients into a food processor with a steel blade. Pulse until everything is blended and chopped. If you are using preground turkey, mix all the ingredients well.
2. Make 12 patties and fry until well done in a nonstick skillet with olive oil.

Nutritional Analysis (per serving)

128	Calories
6 gm.	Fat
49.5 mg.	Cholesterol
42%	of Calories from Fat
17.1 gm.	Protein
1.74 gm.	Saturated Fat
0.14 gm.	Fiber
0.5 gm.	Carbohydrate
283.8 mg.	Sodium
22 mg.	Calcium

Country Sausage Gravy

Serves 4

I grew up with homemade biscuits and gravy. Here is a version that is low in fat and tastes great served over biscuits or rice.

½ pound Turkey Sausage (page 154)
3 tablespoons flour
2 cups 1% milk
Freshly ground black pepper to taste

1. Spray a nonstick skillet with oil, break up sausage, and brown well.
2. Mix flour, milk, and pepper. Pour into skillet with sausage. Bring mixture to a boil, stirring constantly until thick.

Nutritional Analysis (per serving)

133	Calories
4.3 gm.	Fat
29.7 mg.	Cholesterol
29%	of Calories from Fat
13.1 gm.	Protein
1.68 gm.	Saturated Fat
0.14 gm.	Fiber
9.9 gm.	Carbohydrate
82.6 mg.	Sodium
159 mg.	Calcium

Turkey Sausage with Peppers and Onions

Serves 6

This recipe is a low-fat descendent of the traditional Italian sausage and peppers. Serve with pasta or rice.

1 pound Turkey Sausage, made into 12 patties (see page 154)

2 medium onions, sliced

2 green bell peppers, seeded and sliced

1 can (28 ounces) diced tomatoes in juice

½ teaspoon crushed red pepper

½ teaspoon Italian Blend*

6 ounces mozzarella cheese, grated

1. In a large nonstick skillet sprayed with nonstick spray, brown the turkey patties on both sides. Remove from skillet, set aside.
2. Add onions and bell pepper to skillet and sauté until soft. Add tomatoes and seasonings. Simmer 15 minutes.
3. Add sausage patties to skillet and cover them with sauce. Cover and simmer 15 minutes. Uncover and cook 10 minutes.
4. Top with mozzarella and serve.

*See the herb blends on page 40.

Nutritional Analysis (per serving)

225	Calories
10.8 gm.	Fat
52.6 mg.	Cholesterol
43%	of Calories from Fat
21 gm.	Protein
5.34 gm.	Saturated Fat
3.55 gm.	Fiber
11.8 gm.	Carbohydrate
195.2 mg.	Sodium
263 mg.	Calcium

Meat

Where's the Beef?

Chateaubriand
Flank Steak à la India
Ground Beef with Rice
Herb-Crusted Lamb
Lamb Kebabs
Lamb Loin
Meatloaf
Moussaka
Pork Jambalaya
Pork with Tomato-Rosemary Sauce
Roast Pork Tenderloin with Sage
Roast Pork with Brown Rice
Simple Beef Stew

Where's the Beef?

In the past few years we have learned that the most important part of nutrition is balance. Years ago a favorite meal might have been a big steak and a salad; now we know a portion size is 3 to 4 ounces, but we can still enjoy red meat in moderation. A thinly sliced tenderloin served with vegetables, a starch, and a salad can be very satisfying. You can extend meat by making kebabs, a casserole, or a stew.

The secret is making the meat you eat special. The Herb-Crusted lamb for Easter dinner, accompanied by Green Beans Almondine and Potato Pie, is filling and well balanced.

Pork, referred to as the other white meat, is very low in fat and lends itself to many recipes. Pork can be used in many ethnic recipes; the flavor is delicate and conducive to the subtle use of herbs and spices that adds to the complexity of a dish. Of course, you can also spice it up with Spanish or Cajun spices.

Chateaubriand

Serves 4

One of the classics that always presents well and is very easy to make.

1 pound beef tenderloin, all fat trimmed off

1 teaspoon tarragon

1 teaspoon freshly ground black pepper

1. Preheat oven to 400°F.
2. Rinse the tenderloin and pat dry. Rub tarragon and pepper on the meat, then pound it in.
3. Roast for 20 minutes or until the desired doneness. Let rest 10 minutes before slicing. (Remember, the meat will continue to cook as it rests.)

Nutritional Analysis (per serving)

181	Calories
8.6 gm.	Fat
77 mg.	Cholesterol
43%	of Calories from Fat
23.8 gm.	Protein
3.22 gm.	Saturated Fat
0.1 gm.	Fiber
0.6 gm.	Carbohydrate
51.5 mg.	Sodium
16 mg.	Calcium

Flank Steak à la India

Serves 6

Serve this with Indian Raisin Rice (page 198) and chutney on the side.

1½ pounds flank steak, trimmed of all fat
½ teaspoon freshly ground black pepper
1 teaspoon cayenne pepper
½ teaspoon allspice
½ teaspoon cumin
½ teaspoon turmeric

1. Rinse the flank steak and pat dry.
2. Mix the herbs and spices and spread over the meat. Use a mallet to pound the herbs into the meat.
3. Spray the meat lightly with olive oil. Either barbecue or broil the steak for 2 to 4 minutes on each side.
4. Let stand 10 minutes, then slice on the diagonal.

Nutritional Analysis (per serving)

228	Calories
9.2 gm.	Fat
90.8 mg.	Cholesterol
36%	of Calories from Fat
33.6 gm.	Protein
3.46 gm.	Saturated Fat
0.17 gm.	Fiber
0.6 gm.	Carbohydrate
72.3 mg.	Sodium
14 mg.	Calcium

Ground Beef with Rice

Serves 4

This is a great way to use up leftovers. You can also replace the beef with chicken or turkey and use just about any leftover vegetable.

½ pound lean ground beef
½ medium onion, chopped
1 green bell pepper, chopped
1½ teaspoons Italian Blend*
1 can (28 ounces) diced tomatoes
3 cups brown rice, cooked

1. In a large nonstick skillet, sauté the ground beef until it is well cooked. Set it aside on a paper plate to drain.
2. Spray a skillet with nonstick spray, and cook the onion and peppers until soft.
3. Add seasonings and tomatoes and simmer 10 minutes, uncovered.
4. Return the beef to the skillet and add rice. Mix well and heat through.

Nutritional Analysis (per serving)

317	Calories
5.8 gm.	Fat
38.5 mg.	Cholesterol
16%	of Calories from Fat
17.8 gm.	Protein
1.93 gm.	Saturated Fat
6.83 gm.	Fiber
48.9 gm.	Carbohydrate
60.9 mg.	Sodium
86 mg.	Calcium

*See the herb blends on page 40.

Herb-Crusted Lamb

Serves 6

A perfect dish for Easter or any special occasion. When serving, make sure everyone has some of the mustard crust. Green Beans Almondine (page 179) and Potato Pie (page 193) are the perfect complements.

½ cup dry bread crumbs
1 clove garlic, minced
½ teaspoon dry thyme
¼ teaspoon freshly ground black pepper
4 tablespoons Dijon mustard
2 pounds rack of lamb (ask your butcher to trim off all the fat)

1. Preheat the oven to 400°F.
2. Combine all ingredients (except lamb) and rub over the lamb.
3. Roast lamb for 45 to 60 minutes, to desired doneness.

Nutritional Analysis (per serving)

294	Calories
17.4 gm.	Fat
94.2 mg.	Cholesterol
53%	of Calories from Fat
27.2 gm.	Protein
7.06 gm.	Saturated Fat
0.54 gm.	Fiber
5.5 gm.	Carbohydrate
240.6 mg.	Sodium
28 mg.	Calcium

Lamb Kebabs

Serves 4

They are colorful, healthful, and taste great.

1 recipe marinade (page 130)
1 pound very lean lamb, cubed
12 mushrooms, cleaned
1 green bell pepper, cut into 6 cubes
1 red bell pepper, cut into 6 cubes
12 pearl onions, or 12 slices of onion
4 skewers

1. Put the lamb marinade into a lock-top plastic bag with the cubed lamb. Refrigerate for several hours, turning occasionally.

2. Alternate the ingredients on the skewer, starting and finishing with mushrooms.

3. Barbecue or broil, using the marinade to baste the kebabs while cooking.

Nutritional Analysis (per serving)

357	Calories
14.4 gm.	Fat
70.7 mg.	Cholesterol
36%	of Calories from Fat
24.6 gm.	Protein
5.5 gm.	Saturated Fat
6.63 gm.	Fiber
34.3 gm.	Carbohydrate
62.8 mg.	Sodium
85 mg.	Calcium

Lamb Loin

Serves 4

This is particularly tasty barbecued on the outside grill.

2 tablespoons fresh lemon juice
2 cloves garlic, minced
1½ teaspoons fresh oregano (½ teaspoon dried)
1½ teaspoons fresh thyme (½ teaspoon dry)
½ teaspoon freshly ground black pepper
1½ pounds loin of lamb, boned, trimmed, and tied

1. Mix all ingredients (except lamb) together and rub on the lamb. Wrap tightly and refrigerate 4 to 5 hours.
2. Grill or roast in a 350°F oven to desired doneness. Medium rare is about 11 minutes on each side.

Nutritional Analysis (per serving)

301	Calories
18.9 gm.	Fat
106 mg.	Cholesterol
57%	of Calories from Fat
29.4 gm.	Protein
7.88 gm.	Saturated Fat
0.21 gm.	Fiber
1.6 gm.	Carbohydrate
76 mg.	Sodium
23 mg.	Calcium

Meatloaf

Serves 6

If you don't have a special meatloaf pan—one with holes that fits into another pan so the fat can drip out—take a foil loaf pan and poke holes in the bottom of it. Make a few balls of foil and put them in the bottom of a regular loaf pan. Place the foil pan inside the regular pan for baking. You can also form the meatloaf in a rectangular mound and bake it in a large baking dish.

1 pound lean ground beef
1 pound ground turkey
¾ cup egg substitute (or 3 eggs)
1 cup rolled oats (oatmeal)
½ cup ketchup
2 teaspoons dry mustard
½ teaspoon ground black pepper
½ teaspoon Italian Blend*
½ teaspoon paprika
1 medium onion, diced
1 green pepper, diced
1 red bell pepper, diced

TOPPING
¼ cup ketchup
¼ cup Dijon mustard

1. Preheat oven to 350°F.
2. Combine all ingredients except for the topping. Put into a meatloaf pan that has been sprayed with nonstick spray.
3. Mix the ketchup and mustard together and spread over the top of the meatloaf.
4. Bake for 2 hours.

*See the herb blends on page 40.

Nutritional Analysis (per serving)

331	Calories
13.7 gm.	Fat
189.5 mg.	Cholesterol
37%	of Calories from Fat
33.1 gm.	Protein
4.25 gm.	Saturated Fat
2.57 gm.	Fiber
18.9 gm.	Carbohydrate
684.5 mg.	Sodium
62 mg.	Calcium

Moussaka

Serves 8

Moussaka is a traditional Greek dish usually prepared with ground lamb (you can use lean ground beef or turkey) and covered with a thick custard sauce. I've replaced the custard with a lovely nontraditional topping—and no one will guess its low-fat secret. It brings the dish to within healthy guidelines. Moussaka takes more time to prepare, but it is well worth it. The leftovers only get better after it sets in the refrigerator. When preparing Moussaka for a dinner party, make it a day ahead, except for the topping. That gives all the flavors a chance to marry, and the topping is fresh!

2 large eggplants, cut into ¼-inch round slices
2 teaspoons garlic-infused olive oil
2 medium onions, chopped
1 pound lean ground lamb
1 can (27 ounces) diced tomatoes in juice
½ cup chopped parsley
¼ teaspoon ground cinnamon
½ teaspoon freshly ground nutmeg
1 tablespoon fresh oregano (½ teaspoon dried)
1 cup plain yogurt

TOPPING
3 cups 1% milk
2 teaspoons butter
½ teaspoon ground white pepper
¾ cup Cream of Rice cereal
¼ cup Parmesan cheese

Nutritional Analysis
(per serving)

231	Calories
11.2 gm.	Fat
46.1 mg.	Cholesterol
44%	of Calories from Fat
15.9 gm.	Protein
4.89 gm.	Saturated Fat
3.05 gm.	Fiber
17.4 gm.	Carbohydrate
160.7 mg.	Sodium
208 mg.	Calcium

1. Lightly salt eggplant slices and let set for 45 minutes.
2. Preheat oven to 400°F.

(continued on next page)

3. Rinse eggplant and pat dry. Place on cookie sheet that has been sprayed lightly with olive oil. Lightly spray the tops of the eggplant with olive oil. Bake for 15 minutes, remove from the oven, and reduce the temperature to 350°F.
4. Heat olive oil and sauté onion until soft; remove onion from pan and set aside.
5. Put the ground lamb into the skillet and cook until done. Add onions, tomatoes, parsley, cinnamon, nutmeg, and oregano. Mix well and simmer uncovered for 30 minutes.
6. Stir in yogurt.
7. Spray baking pan with olive oil and layer the eggplant on the bottom of the pan. Spread the meat mixture on top of the eggplant.
8. Bake at 350°F for 45 minutes. Let stand 10 minutes before serving.
9. To make the topping, put milk, butter, and white pepper into a saucepan and bring to a boil. Slowly add Cream of Rice cereal, stirring constantly. Add Parmesan cheese. When sauce is thick, pour over the Moussaka.

Pork Jambalaya

Serves 6

A Cajun delight that is a one-dish meal, Jambalaya can be made in a large rice cooker, which turns off automatically when done and keeps the food warm.

1 pound lean pork loin, cubed
2 medium onions, chopped
1½ cup celery chopped
1 clove garlic, chopped
½ teaspoon ground white pepper
½ teaspoon ground black pepper
1 teaspoon dry mustard
½ teaspoon ground cayenne pepper
½ teaspoon cumin
½ teaspoon ground thyme
2 cups short-grain brown rice
4 cups chicken broth
2 bay leaves

1. In a large Dutch oven sprayed with olive oil, brown the pork cubes. Take out and set aside.
2. Put onions, celery, and garlic into pan and cook until soft.
3. Add seasonings to mixture and sauté 30 seconds.
4. Return pork to pan, add rice, chicken broth, and bay leaves. Bring to a boil.
5. Cover and reduce to a simmer. Cook until all liquid is absorbed, or place in oven at 350°F for 1 hour. Remove the bay leaves before serving.

Nutritional Analysis
(per serving)

227	Calories
6.3 gm.	Fat
45.6 mg.	Cholesterol
25%	of Calories from Fat
19.9 gm.	Protein
2.14 gm.	Saturated Fat
2.43 gm.	Fiber
22 gm.	Carbohydrate
584.3 mg.	Sodium
58 mg.	Calcium

Pork with Tomato-Rosemary Sauce

Serves 4

Serve over wide noodles or rice. For more of a cream sauce, finish with 1 cup of light sour cream.

1 pound pork tenderloin, trimmed, cut into 8 slices
2 medium onions, chopped
1½ cups tomato juice
1 teaspoon chopped fresh rosemary (½ teaspoon dried)
⅛ teaspoon cayenne pepper
2 tablespoons flour, mixed in ¼ cup water
Chopped fresh parsley to garnish

1. In a heated skillet sprayed with olive oil, brown pork slices on both sides and remove from pan.
2. Add onions and cook until soft; return pork to skillet.
3. Add tomato juice and seasonings. Cover and simmer until meat is done, about 20 to 25 minutes.
4. Uncover and add flour mixture. Stir until thick. Serve garnished with chopped fresh parsley.

Nutritional Analysis (per serving)

244	Calories
7 gm.	Fat
88.4 mg.	Cholesterol
26%	of Calories from Fat
32.9 gm.	Protein
2.43 gm.	Saturated Fat
1.76 gm.	Fiber
11.5 gm.	Carbohydrate
393.6 mg.	Sodium
35 mg.	Calcium

Roast Pork Tenderloin with Sage

Serves 6

The best plastic bags to use are the ones with "zippers." They hold the seal and are easy to use.

½ cup dry red wine
3 tablespoons minced green onions
2 tablespoons minced fresh sage leaves (2 teaspoons dried)
1 tablespoon minced fresh parsley (1 teaspoon dried)
½ teaspoon dried whole thyme
½ teaspoon white pepper
3 pork tenderloins, about ¾ pound each
Fresh sage sprigs and parsley for garnish

1. Combine the first six ingredients in a lock-top plastic bag. Seal bag and shake well to mix.
2. Trim all fat from the pork. Add pork to the bag, seal, and shake until pork is well coated with mixture.
3. Marinate pork in the bag in the refrigerator for 8 hours. Turn the bag occasionally.
4. Preheat oven to 400°F.
5. Remove the pork from the marinade and place the marinade in a small saucepan. Bring marinade to a boil and cook 5 minutes.
6. Place the pork on a rack in a roasting pan coated with cooking spray. Roast for 45 minutes or until meat is well done (the temperature at the center of the meat should be up to 170°F). Baste the pork frequently with the heated marinade.
7. Transfer the tenderloins to a serving platter. Let meat stand for 10 minutes, then slice thinly diagonally across the grain.
8. Garnish with fresh sage and parsley sprigs and serve.

Nutritional Analysis (per serving)

211	Calories
6.8 gm.	Fat
88.4 mg.	Cholesterol
29%	of Calories from Fat
31.4 gm.	Protein
2.43 gm.	Saturated Fat
0.19 gm.	Fiber
1 gm.	Carbohydrate
64.6 mg.	Sodium
19 mg.	Calcium

Roast Pork with Brown Rice

Serves 4

You can also use thick pork chops or chicken in this recipe.

1 pound tenderloin pork, all fat trimmed off
1 medium onion, sliced
1½ cups brown rice, uncooked
3 cups chicken broth
½ teaspoon freshly ground black pepper
Fresh parsley to garnish

1. Spray Dutch oven with olive oil, heat on high, and brown pork on all sides. Remove pork from pan.
2. Add onion to pan and cook until soft. Add rice and quickly stir for 30 seconds. Turn off heat.
3. Preheat oven to 350°F.
4. Add broth and pepper to rice and stir. Place pork on top of rice, cover and bake for 1 hour.
5. Remove pork, let rest 5 minutes, and then slice. Place sliced pork around the edges of the serving dish with the rice in the center. Garnish with fresh parsley.

Nutritional Analysis (per serving)

391	Calories
8.1 gm.	Fat
88.4 mg.	Cholesterol
19%	of Calories from Fat
37.2 gm.	Protein
2.84 gm.	Saturated Fat
2.96 gm.	Fiber
39.9 gm.	Carbohydrate
656.7 mg.	Sodium
42 mg.	Calcium

Simple Beef Stew

Serves 8

This is as easy as it gets for beef stew! Browning the meat first gives the stew a much richer flavor. This stew can also be made with chicken or pork. Try cooking it in a slow cooker, all day on low.

3 tablespoons flour with ½ teaspoon salt and ½ teaspoon fresh ground black pepper mixed in

1½ pounds very lean beef, cut in large cubes

1 tablespoon garlic olive oil

½ cup red wine or dry vermouth

1 can tomato soup

1½ cups beef stock

1½ teaspoons Italian Blend*

2 packages (16 ounces) frozen stew vegetables

½ cup barley

1. Preheat oven to 300°F.
2. Put flour into a plastic bag, add meat cubes, and shake until the meat is well coated.
3. Heat olive oil in a skillet and brown the meat well. Remove meat to a casserole dish.
4. Pour the wine into the skillet to deglaze. Mix in tomato soup and beef broth, add herbs, and pour over the meat.
5. Mix frozen vegetables and barley into meat, cover, and bake for 4 hours.

Nutritional Analysis (per serving)

329	Calories
11.5 gm.	Fat
60.8 mg.	Cholesterol
32%	of Calories from Fat
27.5 gm.	Protein
4.1 gm.	Saturated Fat
3.81 gm.	Fiber
25.5 gm.	Carbohydrate
8574.5 mg.	Sodium
54 mg.	Calcium

---

*See the herb blends on page 40.

Vegetables

Eat Your Vegetables

Artichokes
Brussels Sprouts
Dilled Carrots
Green Beans
Green Beans Almondine
Green Beans Vinaigrette
Julienne Carrots and Zucchini
Lemon-Herbed Asparagus
Mushrooms Paprika
New Potatoes and Carrots
Oven-Fried Eggplant
Snow Peas with Mushrooms
Stir-Fried Zucchini and Yellow Squash
Tricolor Peppers

Eat Your Vegetables

When you want to cut down on calories, double up on vegetables. They are wonderful in the summer when you can get them freshly picked at your local farmers' market. Remember, you need to use vegetables quickly—they lose their vitamins the longer they sit. When you cannot get fresh vegetables, frozen can be an excellent alternative. The secret to cooking frozen vegetables is to place them in a strainer and thaw them under cold running water. Then heat them in a little chicken broth or use them as directed in your recipe. To get the most flavor and best texture, heat the vegetables right before serving and just long enough for them to get nice and hot.

You can also buy frozen sliced red, yellow, and green bell peppers, which are nice to keep on hand. Spend some time at the grocery store to see what's in the freezer. I prefer the plainer packaging, because then I can do what I want with the vegetables, and they tend to taste much fresher and have fewer calories. My favorite, when I can find them, are whole green beans. They are tender, sweet, and look very nice on a plate.

Celery sticks and baby carrots should be a staple in any refrigerator. When you get it home from the grocer, cut the celery up, wash the sections in cold water, and put them in a lock-top plastic bag. Do the same with the baby carrots. You'll always have celery and carrots ready for a snack or to use in a recipe. I also keep onions in the refrigerator; my eyes don't sting as much when I cut up cold onions. It's worth a try!

Artichokes

___
Serves 4
___

It always surprises me how many people have never had artichokes. Nothing compares with a fresh artichoke served with a dip of mayonnaise or melted butter and lemon juice. First, dip the leaves into the sauce, then eat the meat at the end of the leaf, working your way to the heart, which is the treasure. Artichokes can be served warm or cold.

4 medium artichokes

1 lemon, cut in fourths

2 cups water

½ teaspoon Basic Herb Blend*

1. Cut the stems and tops off of the artichokes. Pull off all the small leaves. Trim the leaves with kitchen shears.
2. Rub the cut parts of the artichokes with lemon to keep them from turning brown.
3. In a Dutch oven bring the water to a boil, add the lemons (squeezing them into the water first) and the Basic Herb Blend.
4. Place artichokes into boiling water, reduce to simmer, and cover, leaving a steaming vent. Cook for 45 minutes or until the bottom is tender. Or put them in a steamer and steam until tender.

Nutritional Analysis
(per serving)

66	Calories
0.3 gm.	Fat
0.0 mg.	Cholesterol
3%	of Calories from Fat
4.5 gm.	Protein
0.06 gm.	Saturated Fat
6.72 gm.	Fiber
15 gm.	Carbohydrate
128.8 mg.	Sodium
104 mg.	Calcium

___
*See the herb blends on page 40.

Brussels Sprouts

Serves 4

Brussels sprouts are tasty and low in calories. After you steam them, you can hold and reheat them in the herbed butter sauce just before serving. They are also good cold.

20 medium Brussels sprouts
1 tablespoon butter
½ teaspoon Basic Herb Blend*

1. Steam Brussels sprouts until they are tender, and then cut in half.
2. In a medium-sized saucepan, melt butter, and add the Basic Blend.
3. Place Brussels sprouts in the saucepan and stir until they are covered with herb butter and heated through.

*See the herb blends on page 40.

Nutritional Analysis (per serving)

73	Calories
3.7 gm.	Fat
8.2 mg.	Cholesterol
45%	of Calories from Fat
3 gm.	Protein
2.02 gm.	Saturated Fat
5.06 gm.	Fiber
10.3 gm.	Carbohydrate
55.8 mg.	Sodium
45 mg.	Calcium

Dilled Carrots

Serves 6

These are good just to keep around as a snack or to use as a garnish on a salad.

1 pound baby carrots
½ cup white wine vinegar
½ cup water
1 teaspoon dillweed
½ teaspoon celery seed

1. Wash carrots.
2. Place all ingredients into a saucepan and bring to a boil. Reduce to simmer, cover, and cook 15 minutes.
3. Drain carrots and place in a covered container in the refrigerator until ready to use.

Nutritional Analysis (per serving)

56	Calories
0.3 gm.	Fat
0.0 mg.	Cholesterol
4%	of Calories from Fat
1.3 gm.	Protein
0.04 gm.	Saturated Fat
3.87 gm.	Fiber
13.5 gm.	Carbohydrate
44.4 mg.	Sodium
47 mg.	Calcium

Green Beans

Serves 4

Green beans are always a perfect standby. They add good color to a meal and everyone loves them. Try to keep some whole green beans in your freezer at all times.

2 cups green beans
¼ cup chicken broth
½ teaspoon minced garlic or shallots

1. If you are using fresh green beans, first stem them. Steam until they are crisp-tender. If you are using frozen, run them under cold water to thaw them before you use them.
2. Heat the chicken broth and garlic in a skillet. Add the green beans, toss to coat, and heat well.

Nutritional Analysis
(per serving)

25	Calories
0.2 gm.	Fat
0.0 mg.	Cholesterol
8%	of Calories from Fat
1.4 gm.	Protein
0.06 gm.	Saturated Fat
0.87 gm.	Fiber
5.3 gm.	Carbohydrate
50.9 mg.	Sodium
32 mg.	Calcium

Green Beans Almondine

Serves 4

The chicken broth will bring out the flavor in the green beans without adding any fat. Toasting the almonds intensifies the flavor so you do not need as many to achieve the taste!

1 pound fresh green beans, washed and trimmed
3 tablespoons chicken broth
2 tablespoons slivered almonds, lightly toasted
Freshly ground black pepper

1. If you are using fresh beans, steam them until crisp-tender, then plunge them into cold water to stop cooking, drain well. If you are using frozen, thaw them first by running cold water over them.
2. Heat chicken broth in a skillet. Add green beans and cook until heated through. Add toasted almonds and serve.

Nutritional Analysis (per serving)

48	Calories
2.4 gm.	Fat
0.0 mg.	Cholesterol
44%	of Calories from Fat
2.1 gm.	Protein
0.26 gm.	Saturated Fat
1.33 gm.	Fiber
6.3 gm.	Carbohydrate
63.6 mg.	Sodium
43 mg.	Calcium

Green Beans Vinaigrette

Serves 6

This dish can double as a vegetable and a salad. It is an excellent choice for a buffet; line a salad bowl with lettuce (red leaf works best) and put the marinated green beans in the center.

1 tablespoon Dijon mustard
3 tablespoons herb vinegar
2 teaspoons olive oil
¼ cup water
1 tablespoon fresh lemon juice
½ teaspoon Italian Blend*
½ teaspoon sugar
Freshly ground pepper to taste
1½ pounds fresh green beans, trimmed
6 large lettuce cups for garnish

1. Combine everything except beans and lettuce, and set aside.
2. Steam beans until just tender. Place into a bowl and pour the vinaigrette over them. Mix well, cover, and refrigerate for at least 2 hours.
3. Serve in large lettuce cups.

Nutritional Analysis (per serving)

61	Calories
2.2 gm.	Fat
0.0 mg.	Cholesterol
33%	of Calories from Fat
2.4 gm.	Protein
0.32 gm.	Saturated Fat
1.66 gm.	Fiber
10.3 gm.	Carbohydrate
36.5 mg.	Sodium
61 mg.	Calcium

*See the herb blends on page 40.

Julienne Carrots and Zucchini

___
Serves 4
___

If you have a mandoline (French vegetable slicer), use it to make the julienne cuts. You can do it in a snap and you don't need to blanch the carrots first. The thin slices cook very quickly and present beautifully.

2 medium carrots
2 teaspoons butter
½ teaspoon dillweed
2 medium zucchini

1. Steam carrots until crisp-tender; this makes them easier to cut. Cut carrots and zucchini into julienne pieces.
2. Melt butter in a medium skillet. Add dillweed, carrots, and zucchini, and gently toss.
3. Turn off the heat, cover, and let stand 2 to 3 minutes before serving.

Nutritional Analysis (per serving)

52	Calories
2.4 gm.	Fat
5.5 mg.	Cholesterol
41%	of Calories from Fat
1.2 gm.	Protein
1.33 gm.	Saturated Fat
2.42 gm.	Fiber
7.6 gm.	Carbohydrate
34.4 mg.	Sodium
37 mg.	Calcium

Lemon-Herbed Asparagus

Serves 6

Asparagus always seems to make a meal special. If you want to use only one pan, you can first put the asparagus in the skillet with 1 inch of water. Bring to a boil, cover, and steam 5 minutes. Remove the asparagus, bring the liquid to a boil, and reduce it to almost nothing. Then heat your herbs and butter right in the skillet. That way you keep the vitamins and minerals that were left in the cooking liquid.

1 pound asparagus, rinsed and trimmed
2 teaspoons butter
¼ teaspoon basil
¼ teaspoon oregano
2 teaspoons fresh lemon juice

1. Steam asparagus in steamer for 5 minutes.
2. In a skillet heat butter, basil, oregano, and lemon juice on low.
3. Add asparagus and toss gently right before serving.

Nutritional Analysis
(per serving)

31	Calories
1.6 gm.	Fat
3.6 mg.	Cholesterol
46%	of Calories from Fat
2 gm.	Protein
0.89 gm.	Saturated Fat
1.3 gm.	Fiber
3.3 gm.	Carbohydrate
22.1 mg.	Sodium
17 mg.	Calcium

Mushrooms Paprika

Serves 4

This can be served over rice as a side dish or as the sauce over chicken or pork. The lemon juice keeps the mushrooms white.

1 tablespoon shallots, finely chopped
1 pound mushrooms, sliced
1 teaspoon fresh lemon juice
2 teaspoons paprika
1/8 teaspoon ground red pepper
1/2 cup light sour cream

1. In a skillet sprayed with olive oil, sauté shallots until soft. Add mushrooms and lemon juice, cover, and simmer for 3 minutes.
2. Remove cover and sauté until all liquid is absorbed.
3. Mix paprika and red pepper into sour cream. Add to mushrooms and heat (do not boil).

Nutritional Analysis (per serving)

43	Calories
3.3 gm.	Fat
6.4 mg.	Cholesterol
69%	of Calories from Fat
1.4 gm.	Protein
1.92 gm.	Saturated Fat
0.68 gm.	Fiber
2.9 gm.	Carbohydrate
9.5 mg.	Sodium
21 mg.	Calcium

New Potatoes and Carrots

Serves 4

If you can not find tiny new potatoes, buy larger ones and cut them in half or in fourths depending on the size.

12 whole tiny new potatoes, washed

4 medium carrots, washed and cut in 1-inch slices on the diagonal

2 tablespoons butter

1 teaspoon dillweed

1. Steam potatoes and carrots, covered, for 20 minutes or until tender.
2. Melt butter and dillweed in a skillet.
3. Put potatoes and carrots into a serving dish and toss with butter and dill.

Nutritional Analysis (per serving)

363	Calories
6.6 gm.	Fat
16.4 mg.	Cholesterol
16%	of Calories from Fat
6.8 gm.	Protein
3.89 gm.	Saturated Fat
8.72 gm.	Fiber
71.7 gm.	Carbohydrate
100.6 mg.	Sodium
43 mg.	Calcium

Oven-Fried Eggplant

Serves 6

Here is a great way to get tender, crispy eggplant that is not full of oil.

1 medium to large eggplant, cut in ¼-inch rounds
2 cups plain bread crumbs
½ teaspoon dry oregano
½ teaspoon garlic powder
½ teaspoon ground black pepper
½ cup all-purpose flour
2 eggs, slightly beaten

1. Preheat oven to 350°F.
2. In a large pot of boiling salted water, blanch eggplant for 3 minutes in small batches. Lay on paper towels to dry.
3. Combine bread crumbs, oregano, garlic powder, and pepper.
4. Dip eggplant into flour, then into the eggs, ending with the bread-crumb mixture.
5. Place dipped eggplant on a cookie sheet sprayed with olive oil. Spray eggplant with olive oil.
6. Bake for 15 minutes.

Nutritional Analysis
(per serving)

168	Calories
2.9 gm.	Fat
70.1 mg.	Cholesterol
16%	of Calories from Fat
6.5 gm.	Protein
0.8 gm.	Saturated Fat
1.99 gm.	Fiber
28.4 gm.	Carbohydrate
215.3 mg.	Sodium
44 mg.	Calcium

Snow Peas with Mushrooms

___
Serves 4
___

Snow peas cook very quickly and are best when they are crisp-tender. If you cannot find fresh ones, buy frozen and defrost them by putting them in a strainer under cold running water. Then, put them in the saucepan just long enough to heat through.

¼ cup chicken broth
4 cups sliced mushrooms
2 cups snow peas, trimmed
¼ cup chopped dill, fresh

1. Heat chicken broth in nonstick skillet. Add mushroom and sauté until tender.
2. Add snow peas and dill and stir 30 seconds.
3. Turn off heat, cover, and let stand 2 minutes before serving.

Nutritional Analysis (per serving)

52	Calories
0.5 gm.	Fat
0.0 mg.	Cholesterol
8%	of Calories from Fat
4.2 gm.	Protein
0.09 gm.	Saturated Fat
3.14 gm.	Fiber
8.9 gm.	Carbohydrate
54.9 mg.	Sodium
38 mg.	Calcium

Stir-Fried Zucchini and Yellow Squash

Serves 4

This is a good standby because you can usually always find fresh zucchini and yellow squash in the market. They grow well in a garden if you have a spot in your yard.

2 teaspoons garlic-infused olive oil
½ onion, thinly sliced
2 medium zucchini, sliced on diagonal
2 medium yellow squash, sliced on diagonal
1 tablespoon fresh basil, thinly sliced (1 teaspoon dried)
⅛ teaspoon crushed red pepper
Freshly ground black pepper to taste

1. In a nonstick skillet heat olive oil. Add onion and cook until soft.
2. Add zucchini and yellow squash, and stir until crisp-tender.
3. Add basil and red and black pepper. Serve.

Nutritional Analysis (per serving)

46	Calories
1.5 gm.	Fat
0.0 mg.	Cholesterol
29%	of Calories from Fat
1.8 gm.	Protein
0.23 gm.	Saturated Fat
2.3 gm.	Fiber
8 gm.	Carbohydrate
2.7 mg.	Sodium
55 mg.	Calcium

Tricolor Peppers

Serves 6

This recipe can be used to add both flavor and color to a variety of dishes.

1 green bell pepper, thinly sliced
1 red bell pepper, thinly sliced
1 yellow bell pepper, thinly sliced
1 onion, thinly sliced
½ cup chicken broth
1 teaspoon Vegetable Blend*

1. Spray a large nonstick skillet with olive oil and heat. Add peppers and onion to pan, and cook over medium-high heat for about 1 minute.
2. Add chicken broth and Vegetable Blend to the mixture, cover, and simmer for 10 minutes. Uncover, bring heat to medium high, and cook until all liquid is absorbed.

Nutritional Analysis (per serving)

23	Calories
0.2 gm.	Fat
0.0 mg.	Cholesterol
8%	of Calories from Fat
0.9 gm.	Protein
0.05 gm.	Saturated Fat
1.02 gm.	Fiber
5.2 gm.	Carbohydrate
67 mg.	Sodium
12 mg.	Calcium

*See the herb blends on page 40.

Potatoes/Orzo/Rice

On the Side

POTATOES

Mashed
Oven French Fries
Potato Pie

ORZO

Greek-Style Orzo
Orzo with Parmesan and Basil

RICE

Date Risotto
Fried Rice
Indian Raisin Rice
Lemon Rice
Wild-Brown Rice

On the Side

Potatoes—mashed, roasted, or fried—are always a favorite at the dinner table. On their own, potatoes are low in calories and high in vitamins and minerals. I've found the secret to a tasty roasted potato or oven fries is to spray them lightly with a good olive oil. I prefer to use the infused olive oils to give them additional flavor. This is where the Misto, one of my favorite new toys, comes in handy. Grind some fresh black pepper or almost any herb on the potatoes to add yet another dimension to the flavor.

Orzo is a bit unusual. It looks like large rice grains, but it is actually pasta. This uniqueness makes it fun to serve.

As long as you keep potatoes, rice, and orzo in your pantry, you'll always be ready for that last-minute company.

Keep a variety of rice in your pantry as it keeps well and is very versatile; however, brown rice can go rancid, so keep it in a cool place or in your refrigerator.

In most of the rice recipes you can interchange the rice. When making a basic rice dish, long-grain white or brown is best. In a risotto or rice pudding Arborio is best as it cooks up creamy as it contains more starch than long-grain rice.

Mashed Potatoes

Serves 6

Everyone loves mashed potatoes. If you keep the skin on, you get more fiber and vitamins.

6 medium potatoes, washed and cubed
1 tablespoon butter
¼ teaspoon ground white pepper
½ teaspoon ground fresh nutmeg
¾ cup 1% milk

1. Place potatoes in a large Dutch oven or saucepan and cover with water. Bring to a boil and simmer until potatoes are tender.
2. Drain off all the water, and mash potatoes with a masher or put them through a ricer.
3. Add the remaining ingredients and whip until smooth.
4. Serve warm.

Nutritional Analysis (per serving)

136	Calories
2.5 gm.	Fat
6.7 mg.	Cholesterol
17%	of Calories from Fat
3.1 gm.	Protein
1.54 gm.	Saturated Fat
2.44 gm.	Fiber
25.9 gm.	Carbohydrate
42.3 mg.	Sodium
49 mg.	Calcium

Oven French Fries

Serves 4

Here is a healthy way to enjoy those forbidden french fries. You can add variety by using different herbs or seasonings. For spicy fries, add cayenne pepper or crushed chili pepper. You may also use a flavor-infused olive oil.

3 large russets, about 1½ pounds
2 teaspoons olive oil
½ teaspoon salt
¼ teaspoon paprika
Freshly ground black pepper.

1. Preheat oven to 450°F. Coat a baking sheet with non-stick spray.
2. Cut potatoes into strips (use a mandoline if you have one).
3. In a large bowl combine olive oil, salt, paprika, and pepper. Add potatoes and toss to coat.
4. Spread the potatoes on a baking sheet and roast about 20 minutes. Loosen and turn the potatoes. Roast 10 to 15 minutes longer or until golden brown.

Nutritional Analysis (per serving)

156	Calories
2.5 gm.	Fat
0.0 mg.	Cholesterol
15%	of Calories from Fat
3.6 gm.	Protein
0.36 gm.	Saturated Fat
2.76 gm.	Fiber
30.7 gm.	Carbohydrate
252.5 mg.	Sodium
13 mg.	Calcium

Potato Pie

Serves 4

Potato pie is a great recipe to play with. You can use any vegetables you have handy, even leftovers. You can also use fresh herbs. Baking times may vary so check for tenderness.

3 medium potatoes, thinly sliced
2 medium carrots, grated
3 green onions, chopped, including green tops
1 cup chicken broth
2 tablespoons chopped parsley

1. Preheat oven to 350°F.
2. Arrange 2 layers of potatoes around a 9-inch pie or quiche pan that has been sprayed with a nonstick coating. Spread the carrots and onions over the potatoes. Layer the remaining potatoes on top.
3. Pour the chicken broth over the top, cover, and bake 1 hour.
4. Uncover and sprinkle parsley over the top. Bake, uncovered, 10 to 15 minutes.
5. Let stand 10 minutes, then cut into pie-shaped wedges.

Nutritional Analysis (per serving)

182	Calories
0.5 gm.	Fat
0.0 mg.	Cholesterol
3%	of Calories from Fat
6.1 gm.	Protein
0.15 gm.	Saturated Fat
5.93 gm.	Fiber
40.7 gm.	Carbohydrate
235 mg.	Sodium
88 mg.	Calcium

Greek-Style Orzo

Serves 6

Orzo is a small pasta that looks like rice. It makes a nice change from the ordinary.

½-ounce dry-packed sun-dried tomatoes
2 cups orzo
½ red onion, chopped
½ red bell pepper, chopped
1 green bell pepper, chopped
2 ounces cumbered feta cheese, crumbled
2 ounces ripe sliced olives
2 tablespoons chopped fresh parsley
1 teaspoon oregano
4 tablespoons wine vinegar

1. Soak sun-dried tomatoes in boiling water for 10 minutes. When cool, chop them up.
2. Cook orzo in a large pot of boiling water until tender; drain, and set aside.
3. In a skillet sprayed with olive oil, sauté onion and peppers until soft.
4. Toss all ingredients with the orzo, reheat, and serve.

Nutritional Analysis
(per serving)

182	Calories
3.5 gm.	Fat
8.3 mg.	Cholesterol
17%	of Calories from Fat
6.6 gm.	Protein
1.61 gm.	Saturated Fat
2.44 gm.	Fiber
31.2 gm.	Carbohydrate
153.3 mg.	Sodium
72 mg.	Calcium

Orzo with Parmesan and Basil

Serves 6

This is delicious served with just-grilled vegetables for a meatless meal.

1½ cups orzo

3 cups low-salt chicken broth

⅓ cup freshly grated Parmesan cheese

5 tablespoons fresh basil, chopped (or 1½ teaspoons dried)

Freshly ground black pepper to taste

1. Place orzo and chicken broth in a saucepan and bring to a boil. Cover and simmer until liquid is absorbed, about 20 minutes.
2. Add Parmesan cheese, basil, and pepper. Mix well, and serve.

Nutritional Analysis (per serving)

128	Calories
1.8 gm.	Fat
2.6 mg.	Cholesterol
13%	of Calories from Fat
6.4 gm.	Protein
0.82 gm.	Saturated Fat
1.57 gm.	Fiber
21.4 gm.	Carbohydrate
453.5 mg.	Sodium
110 mg.	Calcium

Date Risotto

Serves 6

I developed this recipe while working with the California Date Board. Dates are a joy to work with because they are full of flavor as well as vitamins and minerals. The combination of chilies, dates, and walnuts gives this dish an interesting flavor. Try it as a side dish with grilled chicken or pork.

3 tablespoons chili-infused olive oil
1 medium onion, chopped
2 cups arborio rice
½ cup dry vermouth
4 cups chicken broth
½ cup chopped fresh parsley
1 cup date pieces
½ cup chopped walnuts

1. Heat chili oil and sauté onion until soft.
2. Add rice and sauté over medium-high heat until the rice is completely coated with the oil.
3. Add vermouth and stir until absorbed.
4. Add chicken broth 1 cup at a time until the rice absorbs the liquid. (Have extra water ready in case you need it.) Cook until rice is tender, stirring constantly.
5. Add parsley, date nuggets, and walnuts. Stir in well and serve.

Nutritional Analysis (per serving)

456   Calories
14.5 gm.   Fat
0.0 mg.   Cholesterol
29%   of Calories from Fat
8.3 gm.   Protein
1.88 gm.   Saturated Fat
4.91 gm.   Fiber
71.3 gm.   Carbohydrate
535 mg.   Sodium
73 mg.   Calcium

Fried Rice

Serves 4

A quick, easy way to use leftover rice. You can also add leftover vegetables and meats for a one-dish meal.

1 tablespoon butter

2 green onions, chopped, including tops

1 cup corn kernels

1 teaspoon Vegetable Blend*

2 cups cooked brown rice

½ cup egg substitute or 2 whole eggs

1. Heat butter in a skillet. Add onions and cook until soft.
2. Add corn, Vegetable Blend, and rice, and stir until hot.
3. Add egg substitute, stirring until cooked.

Nutritional Analysis (per serving)

265	Calories
7.6 gm.	Fat
8.8 mg.	Cholesterol
27%	of Calories from Fat
8.1 gm.	Protein
2.72 gm.	Saturated Fat
4.56 gm.	Fiber
43.4 gm.	Carbohydrate
103.4 mg.	Sodium
54 mg.	Calcium

*See the herb blends on page 40.

Indian Raisin Rice

Serves 6

A good side dish for anything with an Indian flavor, Indian Raisin Rice can be made with ease in a rice cooker.

3½ cups chicken broth
1½ cups brown rice
1 teaspoon turmeric
½ teaspoon ground cumin
¼ teaspoon ground ginger
⅛ teaspoon cayenne pepper
1 cup raisins

1. Place all ingredients into a saucepan and bring to boil.
2. Cover and simmer until all liquid is absorbed, about 45 minutes.

Nutritional Analysis (per serving)

201 Calories
1.1 gm. Fat
0.0 mg. Cholesterol
5% of Calories from Fat
4.9 gm. Protein
0.35 gm. Saturated Fat
3.04 gm. Fiber
44.4 gm. Carbohydrate
464.7 mg. Sodium
34 mg. Calcium

Lemon Rice

Serves 4

A perfect side dish for chicken or fish, Lemon Rice is especially easy when made in a rice cooker.

2½ cups low-salt chicken broth
1 cup brown rice, uncooked
1 tablespoon fresh lemon zest (2 teaspoons dried)
2 tablespoons fresh dill (2 teaspoons dried)

1. Put broth, rice, lemon zest, and dill into a saucepan, and bring to a boil.
2. Add brown rice.
3. Cover and simmer until liquid is absorbed.

Nutritional Analysis (per serving)

129	Calories
0.9 gm.	Fat
0.0 mg.	Cholesterol
7%	of Calories from Fat
4.3 gm.	Protein
0.32 gm.	Saturated Fat
1.75 gm.	Fiber
25.5 gm.	Carbohydrate
495 mg.	Sodium
31 mg.	Calcium

Wild-Brown Rice

Serves 6

Wild rice and brown rice are an excellent combination, but you may use white rice as well. Wild rice alone can be overpowering. The water chestnuts give this dish a crunchiness that is appealing. For variety, add some toasted pecans. Wild-Brown Rice is delicious as a side dish with Cornish game hens, or as a stuffing for poultry.

1 tablespoon olive oil
1 cup long-grain brown rice
½ cup wild rice, washed
1 medium onion, thinly sliced
3½ cups chicken broth
1 tablespoon soy sauce
1 teaspoon crushed dried thyme
1 can sliced water chestnuts, drained

1. Heat olive oil in a Dutch oven. Add onion and sauté until soft.
2. Add rice and quick stir until rice is light brown.
3. Add the remaining ingredients, and bring to a boil. Reduce to a simmer, cover, and cook until liquid is absorbed (about 45 minutes).

Nutritional Analysis (per serving)

150	Calories
3.2 gm.	Fat
0.0 mg.	Cholesterol
19%	of Calories from Fat
4.7 gm.	Protein
0.59 gm.	Saturated Fat
2.25 gm.	Fiber
25.9 gm.	Carbohydrate
886.3 mg.	Sodium
24 mg.	Calcium

Breakfast

Off to a Great Start!

Baked Oatmeal
Crustless Quiche
Peanut Butter Breakfast Bars
Potato Frittata
Veggie Omelet
Waffles

Off to a Great Start!

Breakfast is the most important meal of the day, but so many people start without it. That's why I developed Peanut Butter Breakfast Bars. They are delicious with coffee or tea and can be carried with you when you're on the run. Baked Oatmeal is also an excellent way to begin the day. Make it the night before and reheat it in the morning, or take out what you want and have it for a couple of mornings. Try making the waffles and freezing them for your own toaster waffle. The Crustless Quiche and other egg dishes might best be served as Sunday brunch. That way you can relax and enjoy them.

Baked Oatmeal

Serves 8

A great, and different, way to have oatmeal. Use any dried fruit you like; chopped nuts are also a nice addition.

2¼ cups oatmeal

½ cup brown sugar

½ cup currants or raisins

1 teaspoon cinnamon

1 teaspoon vanilla

3⅓ cups skim milk or light vanilla soy milk

½ cup egg substitute or 2 eggs

1. Preheat oven to 350°F.
2. Mix all ingredients together and pour into an 8-inch baking dish sprayed with nonstick spray.
3. Bake for 60 minutes.
4. Serve with milk or yogurt.

Nutritional Analysis (per serving)

147	Calories
1.4 gm.	Fat
2 mg.	Cholesterol
8%	of Calories from Fat
7.1 gm.	Protein
0.34 gm.	Saturated Fat
1.81 gm.	Fiber
26.8 gm.	Carbohydrate
186.2 mg.	Sodium
155 mg.	Calcium

Crustless Quiche

Serves 6

Quiche is great as an appetizer, part of a brunch buffet, a luncheon entrée, or a late-night supper.

4 eggs or 8 ounces egg substitute
4 egg whites
½ cup flour
1 teaspoon baking powder
1 teaspoon marjoram
2 cups nonfat cottage cheese
1 cup green chilies, chopped
3 ounces Monterey Jack cheese, grated
3 ounces Mozzarella cheese, grated

1. Preheat oven to 400°F.
2. Put eggs or egg substitutes, egg whites, flour, baking powder, marjoram, and cottage cheese into work bowl of food processor and process with a steel blade until well blended and smooth.
3. Add green chilies and cheese to processor and pulse on and off until just blended.
4. Spray a 9-by-13-inch baking dish with nonstick spray. Pour mixture into baking dish and bake 15 minutes. Lower heat to 350°F and bake 30 minutes longer.
5. Cool 10 minutes and cut into 1-inch squares.

Nutritional Analysis
(per serving)

251	Calories
11.9 gm.	Fat
165.7 mg.	Cholesterol
43%	of Calories from Fat
24.1 gm.	Protein
6.33 gm.	Saturated Fat
0.29 gm.	Fiber
10.8 gm.	Carbohydrate
604.1 mg.	Sodium
318 mg.	Calcium

Peanut Butter Breakfast Bars

Serves 16

For breakfast on the run, these bars are perfect.

2 cups reduced-fat peanut butter
½ cup sugar
¼ cup molasses
½ cup egg substitute
2 teaspoons vanilla
1 cup nonfat milk
2 cups Reduced Fat Bisquick
3 cups rolled oats

1. Preheat oven to 350°F.
2. Cream together peanut butter, sugar, molasses, egg substitute, and vanilla.
3. Add milk, Bisquick, and rolled oats, and mix well.
4. Pour mixture into a 9-by-13-inch baking pan sprayed with nonstick spray. Bake for 15 to 20 minutes (be careful not to overcook).
5. Cool and cut into 16 bars.

Nutritional Analysis (per serving)

363	Calories
15.2 gm.	Fat
0.4 mg.	Cholesterol
38%	of Calories from Fat
14.1 gm.	Protein
2.9 gm.	Saturated Fat
6.93 gm.	Fiber
45.2 gm.	Carbohydrate
203.5 mg.	Sodium
69 mg.	Calcium

Potato Frittata

Serves 6

Top with low-fat sour cream and salsa for extra zip.

3 medium russet potatoes, thinly sliced
¼ teaspoon paprika
1 teaspoon Italian Blend*
2 ounces Parmesan cheese, freshly grated
4 whole eggs plus 8 egg whites, beaten together
2 tablespoons water

1. Spray a 10-inch skillet with olive-oil spray and heat.
2. Toss potatoes with seasonings, blending well. Put in hot skillet and cook potatoes until they brown on the bottom.
3. Spray the top of the potatoes with olive-oil spray, turn over, and cook until light brown.
4. Combine cheese and eggs, pour over potatoes, and reduce heat to medium. Lift the potatoes and let some of the egg mixture go underneath them.
5. Add 2 tablespoons of water, cover, and reduce heat to low. Cook until eggs are set, about 8 to 10 minutes.

Nutritional Analysis
(per serving)

169	Calories
6.4 gm.	Fat
147.5 mg.	Cholesterol
34%	of Calories from Fat
13.8 gm.	Protein
2.9 gm.	Saturated Fat
1.25 gm.	Fiber
13.5 gm.	Carbohydrate
290.3 mg.	Sodium
156 mg.	Calcium

*See the herb blends on page 40.

Veggie Omelet

Serves 2

You can create a multitude of omelets. Get inspired by looking at the leftovers in your refrigerator!

2 teaspoons butter
1 cup sliced mushrooms
½ yellow onion, chopped
½ green pepper, chopped
½ tomato, chopped
3 egg whites
2 eggs
1 tablespoon cold water
1 teaspoon fresh dillweed (or ½ teaspoon dried)
Freshly ground black pepper to taste
2 tablespoons light sour cream
Sprigs fresh parsley to garnish

1. Heat nonstick omelet pan and put 1 teaspoon butter in. Add mushrooms, onion, and green pepper. Stir-fry one minute, cover, and reduce heat. Simmer for 5 minutes.
2. Uncover pan and bring up the heat. Add tomatoes and sauté until all liquid is absorbed. Remove from skillet and keep warm.
3. Whisk egg whites, eggs, water, and dillweed together until foamy.
4. Place 1 teaspoon of butter in the nonstick omelet pan, and heat to medium high. Pour in eggs and tilt pan to cover bottom evenly. Use a spatula to move the mixture so it cooks evenly.
5. When eggs are cooked, fill one side of the omelet with the chopped vegetables. Lift the other side of the omelet to fold it over the top.
6. Sprinkle with black pepper. Cut in half and move to plate, garnish with 1 tablespoon sour cream, sprinkle with parsley, and serve.

Nutritional Analysis
(per serving)

184	Calories
11.1 gm.	Fat
224.2 mg.	Cholesterol
55%	of Calories from Fat
13.3 gm.	Protein
5.12 gm.	Saturated Fat
1.62 gm.	Fiber
8.1 gm.	Carbohydrate
194.1 mg.	Sodium
52 mg.	Calcium

Waffles

Makes six 8-inch waffles

The cornmeal in these waffles gives them an extra crunch. Serve them with Yogurt Cream (page 228). Or whip up some cottage cheese with a little sugar and vanilla, and top with fruit.

1 2/3 cups Reduced Fat Bisquick
1/3 cup yellow cornmeal
1 1/3 cup 1% milk
1 teaspoon vanilla
1/4 cup egg substitute
2 tablespoons melted butter

1. Heat waffle iron.
2. Mix all ingredients together.
3. Spray the waffle iron with a nonstick spray, and pour the batter into it.
4. Bake until the steaming stops.

Nutritional Analysis (per serving)

134	Calories
5.9 gm.	Fat
13.1 mg.	Cholesterol
39%	of Calories from Fat
4.4 gm.	Protein
3.11 gm.	Saturated Fat
0.56 gm.	Fiber
15.4 gm.	Carbohydrate
219.2 mg.	Sodium
88 mg.	Calcium

Breads

A Loaf of Bread, a Bottle of Wine, and Thou

Apricot Nut Bread
Cheddar-Beer Bread
Cheese Monkey Bread
Chocolate Chip and Walnut Scones
Cinnamon Nut Bread
Cinnamon Rolls
Green Chili Cornbread
Herbed Spoon Bread
Nut Bread
Poppy Seed Rolls
Pumpkin Muffins
Savory Biscuits
Scones

A Loaf of Bread, a Bottle of Wine, and Thou

I have always thought the saying, "A loaf of bread and bottle of wine and thou," was wonderful. And it is true. A lovely afternoon can be spent with such a simple picnic.

You will also find great teacakes on the following pages. The Reduced Fat Bisquick comes in very handy here; its use makes the recipes quick and easy, as well as pleasing to the palate. I also use the canned Pillsbury French loaf to create a couple of fast recipes. Finally, I've thrown in a couple of twists to cornbread with a chili cornbread and a spoon bread. Both will delight and are perfect with soup, stew, and chili.

Apricot Nut Bread

Serves 12

A perfect snack to have with midmorning coffee or afternoon tea.

2 eggs or ½ cup egg substitute
2 cups low-fat buttermilk
1 teaspoon vanilla
3 cups sifted all-purpose flour
1 cup oat bran
½ cup sugar
¾ cup dried apricots, chopped
½ teaspoon salt
½ teaspoon allspice
2 teaspoons baking powder
½ cup chopped walnuts

1. Preheat oven to 325°F.
2. Beat eggs, buttermilk, and vanilla together. Set aside.
3. Combine dry ingredients, and mix into the eggs and buttermilk. Add the chopped walnuts.
4. Pour the mixture into a 10-inch Bundt pan or a large loaf pan sprayed with nonstick spray. Bake 45 minutes to 1 hour.
5. Turn out onto wire rack and let cool before serving.

Nutritional Analysis (per serving)

225	Calories
4.8 gm.	Fat
36.5 mg.	Cholesterol
19%	of Calories from Fat
6.8 gm.	Protein
0.85 gm.	Saturated Fat
1.92 gm.	Fiber
39.1 gm.	Carbohydrate
208.7 mg.	Sodium
107 mg.	Calcium

Cheddar-Beer Bread

Serves 8

Cheddar-Beer Bread is a great snack to serve during the Sunday football game. Add a pot of chili and your favorite brew, and you have the whole meal covered.

3 cups unbleached flour
1 tablespoon baking powder
1 teaspoon Vegetable Blend*
1 tablespoon sugar
1 cup grated sharp cheddar cheese
1 cup grated mozzarella cheese
½ cup diced onion
1 can (12 ounces) light beer

1. Preheat oven to 350°F.
2. Combine all ingredients except beer and mix well.
3. Slowly add beer until all is well blended.
4. Spoon mixture into a loaf pan sprayed with nonstick spray.
5. Smooth down mixture until even. Bake 1 hour.

Nutritional Analysis (per serving)

283	Calories
8.3 gm.	Fat
24.5 mg.	Cholesterol
26%	of Calories from Fat
12 gm.	Protein
5.09 gm.	Saturated Fat
1.33 gm.	Fiber
36.8 gm.	Carbohydrate
324.2 mg.	Sodium
310 mg.	Calcium

*See the herb blends on page 40.

Cheese Monkey Bread

Serves 6

This is a quick and easy way to make great-tasting bread that will complement any dish.

1 can (11 ounces) Pillsbury French Loaf
½ cup low-fat cheddar cheese, grated
2 teaspoons paprika
¼ teaspoon cayenne pepper

1. Preheat oven to 350°F.
2. Cut French bread dough in half and then into 20 pieces.
3. Place half the bread pieces in a loaf pan sprayed with nonstick spray.
4. Sprinkle the top with cheese, paprika, and cayenne pepper. Top with the remaining bread pieces.
5. Spray lightly with olive-oil spray. Bake for 30 minutes. When slightly cooled, turn the bread out onto a platter. Pull apart to serve.

Nutritional Analysis (per serving)

143	Calories
2.4 gm.	Fat
2 mg.	Cholesterol
15%	of Calories from Fat
6.5 gm.	Protein
1.27 gm.	Saturated Fat
0.99 gm.	Fiber
23.1 gm.	Carbohydrate
382.4 mg.	Sodium
67 mg.	Calcium

Chocolate Chip and Walnut Scones

Serves 8

These quick, easy scones are great to include in your holiday gift basket or to serve with afternoon tea.

2¼ cup Reduced Fat Bisquick
1 tablespoon sugar
½ teaspoon vanilla extract
¼ teaspoon almond extract
½ cup mini semisweet chocolate chips
½ cup chopped walnuts
⅔ cup 1% milk

1. Preheat oven to 425°F.
2. Mix ingredients together.
3. Dust working surface with Bisquick and knead dough 10 times. Roll out dough until it's ½ inch thick, and cut into triangles.
4. Bake on cookie sheet for 8 to 11 minutes or until golden brown.

Nutritional Analysis (per serving)

158	Calories
9 gm.	Fat
0.7 mg.	Cholesterol
51%	of Calories from Fat
3 gm.	Protein
2.59 gm.	Saturated Fat
1.19 gm.	Fiber
18.2 gm.	Carbohydrate
151.9 mg.	Sodium
50 mg.	Calcium

Cinnamon Nut Bread

Serves 6

A quick and easy way to make great-tasting breakfast bread that is sure to please your family and weekend guests.

1 can (11 ounces) Pillsbury French loaf
½ cup walnuts
¼ cup brown sugar
1 teaspoon ground cinnamon

1. Preheat oven to 350°F.
2. Cut French bread dough in half, then into 20 pieces. Place half the bread pieces in a ring pan sprayed well with nonstick spray.
3. Place walnuts, brown sugar, and cinnamon in a mini-chopper and grind until all is finely chopped and mixed together.
4. Sprinkle the top of the bread pieces with the nut mixture and top with remaining bread pieces. Spray lightly with olive oil.
5. Bake for 30 minutes. When slightly cooled, turn bread out onto platter. Pull pieces apart for a tasty treat.

Nutritional Analysis (per serving)

227	Calories
8.1 gm.	Fat
0.0 mg.	Cholesterol
32%	of Calories from Fat
5.7 gm.	Protein
1.42 gm.	Saturated Fat
1.42 gm.	Fiber
33.6 gm.	Carbohydrate
329.7 mg.	Sodium
22 mg.	Calcium

Cinnamon Rolls

Serves 6

These rolls are quite interesting. They are very moist, and the cottage cheese gives them added protein and calcium.

1 cup cottage cheese
2 tablespoons sugar
1 teaspoon vanilla
1½ cup Reduced Fat Bisquick
¼ cup brown sugar
1 tablespoon cinnamon
¼ cup chopped walnuts

1. Preheat oven to 450°F.
2. In a food processor with a steel blade, process cottage cheese until smooth.
3. Add sugar and vanilla to cottage cheese, and pulse to blend. Add Bisquick and pulse to blend.
4. Turn mixture out onto a board sprinkled with Bisquick. Knead the dough, and roll it out into a rectangle ½ inch thick.
5. Mix the brown sugar, cinnamon, and walnuts together, and spread over the dough.
6. Roll up the dough, and place it on a cookie sheet lined with foil; bake 10 minutes. Let cool 15 minutes before cutting.

Nutritional Analysis (per serving)

153	Calories
4.4 gm.	Fat
1.7 mg.	Cholesterol
26%	of Calories from Fat
6.2 gm.	Protein
0.69 gm.	Saturated Fat
0.68 gm.	Fiber
22.7 gm.	Carbohydrate
282.3 mg.	Sodium
63 mg.	Calcium

Green Chili Cornbread

___

Serves 8
___

The green chilies and whole corn kernels put some excitement into this cornbread. It's perfect with bean soup or chili.

1 cup hot water
½ cup sun-dried tomatoes, dry packed
1 cup all-purpose flour
1 cup yellow cornmeal
2 tablespoons sugar
1 teaspoon baking powder
½ teaspoon salt
1 cup low-fat buttermilk
8 ounces egg substitute
1 cup corn kernels
1 can (4 ounces) green chilies

1. Preheat oven to 375°F.
2. Place tomatoes in hot water and let stand 10 minutes, then chop.
3. Combine all dry ingredients in a large bowl.
4. Combine buttermilk and egg substitute, and mix well.
5. Stir the egg-buttermilk mixture into the dry ingredients. Add the corn, tomatoes, and chilies, and stir well.
6. Pour mixture into a 9-inch square baking pan sprayed with nonstick spray. Bake 30 minutes or until done.

Nutritional Analysis
(per serving)

218	Calories
1.9 gm.	Fat
1.4 mg.	Cholesterol
8%	of Calories from Fat
9.9 gm.	Protein
0.47 gm.	Saturated Fat
2.79 gm.	Fiber
40.9 gm.	Carbohydrate
268.2 mg.	Sodium
100 mg.	Calcium

Herbed Spoon Bread

Serves 6

Spoon bread is similar to Italian polenta, but it originated with Native Americans.

2½ cups nonfat milk
1 cup yellow cornmeal
¾ cup egg substitute or 3 eggs
2 tablespoons sugar
1 teaspoon baking powder
2 tablespoons melted butter
1 teaspoon Vegetable Blend*
½ teaspoon salt

1. Heat 2 cups of milk to simmer, and stir in cornmeal. When mixture becomes very thick, remove it from the heat.
2. Preheat oven to 400°F.
3. Combine the remaining ingredients, including the ½ cup of milk, mixing well.
4. Beat the two mixtures together and pour into a 2-quart baking dish sprayed with a nonstick spray.
5. Bake for 45 minutes. Serve at once by spooning onto serving plates.

Nutritional Analysis (per serving)

197	Calories
5.6 gm.	Fat
13 mg.	Cholesterol
26%	of Calories from Fat
9.1 gm.	Protein
2.9 gm.	Saturated Fat
1.26 gm.	Fiber
27.3 gm.	Carbohydrate
380.6 mg.	Sodium
189 mg.	Calcium

*See the herb blends on page 40.

Nut Bread

Serves 12

Nut bread is perfect to serve with a large salad at a luncheon or as a light supper. You can either bake this in one large loaf pan or split it into two smaller loaves, so you can freeze one for later use.

1 egg and 1 egg white
2 cups nonfat milk
1 teaspoon vanilla
3 cups sifted all-purpose flour
4 teaspoons baking powder
1 cup oatmeal
¾ cup sugar
¼ teaspoon salt
1 teaspoon allspice
¼ teaspoon cinnamon
½ cup chopped walnuts

1. Preheat oven to 325°F.
2. Beat eggs, milk, and vanilla together. Set aside.
3. Combine dry ingredients. Mix in milk mixture and chopped walnuts.
4. Pour into large loaf pan sprayed with nonstick spray. Bake for 45 to 60 minutes.
5. Turn onto a wire rack and let cool before serving.

Nutritional Analysis
(per serving)

222	Calories
4.2 gm.	Fat
18.3 mg.	Cholesterol
17%	of Calories from Fat
6.4 gm.	Protein
0.56 gm.	Saturated Fat
1.5 gm.	Fiber
39.6 gm.	Carbohydrate
245.7 mg.	Sodium
149 mg.	Calcium

Poppy Seed Rolls

Serves 6

A quick and lazy way to make dinner rolls. You can use caraway seeds or sesame seeds instead of poppy seeds to vary this recipe.

1 can (11 ounces) Pillsbury French Loaf

2 tablespoons poppy seeds

1. Preheat oven to 350°F.
2. Cut bread into 12 pieces. Roll each piece in poppy seeds.
3. Spray a cookie sheet with nonstick spray or line it with foil (foil makes for an easy cleanup). Place rolls on the cookie sheet and spray lightly with olive oil.
4. Bake until brown, about 10 to 12 minutes.

Nutritional Analysis (per serving)

141	Calories
3 gm.	Fat
0.0 mg.	Cholesterol
19%	of Calories from Fat
4.7 gm.	Protein
0.98 gm.	Saturated Fat
1.34 gm.	Fiber
23.2 gm.	Carbohydrate
325.6 mg.	Sodium
43 mg	Calcium

Pumpkin Muffins

Serves 24

I like using pumpkin in recipes because it adds so much moisture. These muffins make great gifts at the holidays—just wrap them up in a big cloth napkin, or make them part of a holiday basket.

1½ cups unbleached flour
2 teaspoons baking powder
½ cup brown sugar
2 teaspoons cinnamon
¼ teaspoon allspice
¼ teaspoon ginger
¼ teaspoon ground cloves
¼ cup molasses
1 cup canned solid pumpkin
½ cup skim milk
4 ounces egg substitute
½ cup chopped dates
½ cup chopped walnuts

1. Preheat oven to 375°F.
2. Mix all dry ingredients together. Set aside.
3. Combine all liquid ingredients. Add dry ingredients and mix until moistened.
4. Fold in dates and nuts. Spoon into 24 muffin tins sprayed with nonstick spray.
5. Bake for 15 minutes or until done.

Nutritional Analysis (per serving)

93	Calories
1.9 gm.	Fat
0.1 mg.	Cholesterol
19%	of Calories from Fat
2.1 gm.	Protein
0.22 gm.	Saturated Fat
1.06 gm.	Fiber
17.5 gm.	Carbohydrate
51.4 mg.	Sodium
51 mg.	Calcium

Savory Biscuits

Serves 8

When you run out of bread at the last minute, whip these up.

2 cups Reduced Fat Bisquick
1 teaspoon Vegetable Blend*
3/4 cup 1% milk

1. Preheat oven to 450°F.
2. Mix ingredients together.
3. Drop by the spoonful on an ungreased cookie sheet. Or dust a board with Bisquick, turn out the dough and knead it 10 times. Then roll it out to 1/2 inch thick, and cut it into rounds with a cutter.
4. Bake 7 to 9 minutes or until golden brown.

Nutritional Analysis (per serving)

48	Calories
1 gm.	Fat
0.9 mg.	Cholesterol
19%	of Calories from Fat
1.5 gm.	Protein
0.31 gm.	Saturated Fat
0.24 gm.	Fiber
8 gm.	Carbohydrate
136.8 mg.	Sodium
45 mg.	Calcium

*See the herb blends on page 40.

Scones

Serves 8

Scones have become very popular. These are not only very easy to make but also have a lot less fat than the traditional ones. Serve them with Yogurt Cream and jam (page 228).

2¼ cups Reduced Fat Bisquick
1 tablespoon sugar
½ teaspoon vanilla extract
¼ teaspoon almond extract
¼ cup currants
⅔ cup 1% milk

1. Preheat oven to 425°F.
2. Mix ingredients together.
3. Dust a board with Bisquick, turn out the dough and knead it 10 times. Then roll it out to ½ inch thick. Cut into triangles.
4. Place on an ungreased cookie sheet and bake 8 to 11 minutes or until golden brown.

Nutritional Analysis (per serving)

59	Calories
1.1 gm.	Fat
0.7 mg.	Cholesterol
16%	of Calories from Fat
1.5 gm.	Protein
0.29 gm.	Saturated Fat
0.36 gm.	Fiber
10.5 gm.	Carbohydrate
149.9 mg.	Sodium
41 mg.	Calcium

Desserts

Everyone for Dessert

Yogurt Cheese
Yogurt Cream
Apple-Walnut Cake
Basic Sponge Cake
Black Forest Chocolate Cake
Brownies
Chocolate-Chip Sponge Cake
Chocolate Swirl Cheesecake
Dark-and-Spicy Gingerbread
Fruit-Batter Pudding
Lemon Cake
Lemon Cheesecake
Peach Cobbler
Pear Cake
Poppy Seed–Lemon Cake
Pumpkin Cake
Pumpkin Pie
Shortcakes
Spiced Apricots
Tomato-Soup Cake

Everyone for Dessert

Sweets, we love them at any age, and I am certainly at the head of the line! So I'm always trying to come up with a dessert that falls into the guidelines of healthful eating. One of the secrets is to use good extracts such as vanilla, almond, or lemon. Another secret is to be creative. For instance, I use pumpkin purée to give brownies moisture and texture. I use fruit in many of my recipes to add moisture and flavor.

As much as I love dessert, I don't like ones that are too sweet. Cutting down on sugar is just like cutting down on salt. It takes awhile to get used to less salt; then one day foods start tasting too salty. Soon many desserts will be just too sweet.

Yogurt Cheese

Makes 8 ounces

An excellent nonfat replacement for high-fat items like cream cheese, mayonnaise, or sour cream. You can also make it into a sweet cream or spa spice dip.

16 ounces plain nonfat yogurt, without any added gelatin

1. Place yogurt in a yogurt cheese strainer or a colander lined with coffee filters. Put strainer in a large bowl to catch liquid and cover the top. Refrigerate for 18 to 24 hours.
2. Throw out liquid and store Yogurt Cheese in a covered container until ready to use.

Nutritional Analysis (per serving)

34	Calories
0.1 gm.	Fat
1.1 mg.	Cholesterol
3%	of Calories from Fat
3.5 gm.	Protein
0.07 gm.	Saturated Fat
0.0 gm.	Fiber
4.7 gm.	Carbohydrate
46.9 mg.	Sodium
122 mg.	Calcium

Yogurt Cream

Makes 8 servings

A nonfat creamy topping that is excellent on top of fruit desserts.

1 cup Yogurt Cheese (page 227)
½ teaspoon pure vanilla
4 packages Equal or 1 tablespoon sugar

Mix all ingredients together and store in refrigerator in a covered container until ready to use.

Nutritional Analysis (per serving)

37	Calories
0.1 gm.	Fat
1.1 mg.	Cholesterol
3%	of Calories from Fat
3.5 gm.	Protein
0.07 gm.	Saturated Fat
0.0 gm.	Fiber
5.3 gm.	Carbohydrate
48.9 mg.	Sodium
122 mg.	Calcium

Apple-Walnut Cake

Serves 8

This apple cake is good for breakfast, brunch, tea, or dessert!

2 apples, cored and peeled
½ cup walnuts plus ¼ cup chopped walnuts set aside
¾ cup egg substitute
1 can (15 ounces) applesauce
1 teaspoon vanilla
3 cups Reduced Fat Bisquick
1 teaspoon cinnamon
½ cup sugar

1. Preheat oven to 350°F.
2. Put the ingredients, in the order they're listed, into a food processor with a steel blade. Pulse until the mixture is well mixed. Be careful not to mix it too long.
3. Spray a 10-inch Bundt pan with nonstick spray. Sprinkle the reserved walnuts evenly in the bottom, and then spoon the batter over the walnuts.
4. Bake for 45 minutes. Let cool for 10 minutes before unmolding.

Nutritional Analysis (per serving)

244	Calories
9.3 gm.	Fat
0.2 mg.	Cholesterol
34%	of Calories from Fat
5.6 gm.	Protein
1.06 gm.	Saturated Fat
2.05 gm.	Fiber
36.4 gm.	Carbohydrate
229.4 mg.	Sodium
52 mg.	Calcium

Basic Sponge Cake

Serves 8

This is a great all-around recipe to use as a basis for a multitude of creations. The order in which this cake is prepared makes it easier to clean up the kitchen!

5 egg whites
½ cup sugar
2 eggs or 4 ounces egg substitute
2 teaspoons pure vanilla extract
1 teaspoon almond extract
¼ cup water
1 cup flour
1 teaspoon baking powder
¼ teaspoon salt

1. Preheat oven to 350°F.
2. In a clean bowl, beat egg whites, gradually adding ¼ cup sugar until stiff peaks form. Set aside.
3. In a new bowl, beat eggs with ¼ cup sugar, vanilla, almond extract, and water.
4. Sift flour, baking powder, and salt onto the top of egg mixture, and mix in well.
5. Fold 1 cup of the beaten egg whites into the flour mixture and mix to lighten. Pour the batter into remaining beaten egg whites, and gently fold it in.
6. Spray a 10-inch tube pan with nonstick spray, pour the cake batter into the pan, and bake for 35 minutes.
7. Let cool 45 minutes before serving.

Nutritional Analysis
(per serving)

134	Calories
1.5 gm.	Fat
52.6 mg.	Cholesterol
10%	of Calories from Fat
5.2 gm.	Protein
0.43 gm.	Saturated Fat
0.39 gm.	Fiber
23.8 gm.	Carbohydrate
164.5 mg.	Sodium
44 mg.	Calcium

Black Forest Chocolate Cake

Serves 8

An easy way to make an old favorite, this cake is very moist. The Crème de Cassis adds a nice touch—you can drizzle an additional ¼ to ½ cup of it over the cake after you bake it.

1 can (16 ounces) pitted dark cherries, drained
½ cup Crème de Cassis
2½ cups Reduced Fat Bisquick
1 cup Dutch cocoa
1 teaspoon instant coffee granules
1 cup brown sugar
1½ cups low-fat buttermilk
¾ cup egg substitute
1 teaspoon vanilla

1. Marinate cherries in Crème de Cassis for at least 1 hour.
2. Preheat oven to 350°F.
3. Mix dry ingredients together and stir with a balloon whip.
4. Mix the buttermilk, egg substitute and vanilla into the dry ingredients. Fold in the cherries and Crème de Cassis.
5. Spray a tube pan with nonstick spray. Spoon the batter into it. Bake for 30 minutes. Cool for 10 minutes before unmolding.

Nutritional Analysis (per serving)

252	Calories
3.9 gm.	Fat
1.8 mg.	Cholesterol
14%	of Calories from Fat
7.8 gm.	Protein
1.63 gm.	Saturated Fat
4.12 gm.	Fiber
48.7 gm.	Carbohydrate
258.8 mg.	Sodium
127 mg.	Calcium

Brownies

―――――――――
Serves 16
―――――――――

Here's a way to enjoy brownies that aren't loaded with fat and sugar, even though they taste as if they are! You can also stir in ½ cup of chopped walnuts. Or try sprinkling ⅓ cup of finely chopped walnuts into the bottom of the baking pan—when you turn the brownies out, the nuts will be on top. Dutch or European style cocoa powder is darker and richer in taste.

4 ounces egg substitute or 2 eggs
½ cup nonfat milk
1 teaspoon vanilla
1 cup sugar
1 cup solid canned pumpkin
2 cup unbleached flour
2 teaspoons baking powder
1 teaspoon baking soda
½ cup Dutch-style cocoa powder
¾ cup mini semisweet chocolate chips

1. Preheat oven to 350°F.
2. Mix together eggs, milk, vanilla, sugar, and pumpkin.
3. Sift flour, baking powder, baking soda, and cocoa powder into the liquid mixture and mix well. Fold in the chocolate chips.
4. Pour into a tube pan sprayed with nonstick spray, and bake for 40 minutes.
5. Cool and cut into 16 bars.

Nutritional Analysis (per serving)

163	Calories
3.7 gm.	Fat
26.4 mg.	Cholesterol
20%	of Calories from Fat
3.7 gm.	Protein
1.9 gm.	Saturated Fat
2.2 gm.	Fiber
31.6 gm.	Carbohydrate
138.2 mg.	Sodium
57 mg.	Calcium

Chocolate-Chip Sponge Cake

___

Serves 6
___

A very simple variation on the basic sponge cake, it's guaranteed to put big smiles on the faces of chocolate lovers. You can also fold in 1/2 cup of chopped nuts at the end.

5 egg whites
1/2 cup sugar
2 eggs or 4 ounces egg substitute
2 teaspoons pure vanilla extract
1 teaspoon almond extract
1/4 cup water
1 cup flour
1 teaspoon baking powder
1/4 teaspoon salt
1/2 cup mini chocolate chips

1. Preheat oven to 350°F.
2. In a clean bowl, beat egg whites, gradually adding 1/4 cup sugar until stiff peaks form. Set aside.
3. Beat eggs with 1/4 cup sugar, vanilla, almond extract, and water.
4. Sift flour, baking powder, and salt onto the top of the egg mixture. Mix in well.
5. Fold 1 cup of the beaten egg whites into the flour mixture and mix to lighten. Pour batter into the remaining beaten egg whites and fold. Gently fold in chocolate chips.
6. Spray a 10-inch tube pan with a nonstick spray. Pour the cake batter into the pan, and bake for 35 minutes.
7. Let cool 45 minutes before serving.

Nutritional Analysis
(per serving)

175	Calories
4 gm.	Fat
29.8 mg.	Cholesterol
21%	of Calories from Fat
4.7 gm.	Protein
2.1 gm.	Saturated Fat
0.99 gm.	Fiber
30.4 gm.	Carbohydrate
153.9 mg.	Sodium
45 mg.	Calcium

Chocolate Swirl Cheesecake

Serves 8

If you love rich European chocolate, you will love this cheesecake. If you prefer a lighter milk chocolate, use milk chocolate pieces instead of the semisweet and regular cocoa powder instead of the Dutch style. This cheesecake is best if you make it a day before you need it.

CRUST

1 cup graham cracker crumbs
¼ teaspoon cinnamon
¼ teaspoon allspice
3 tablespoons melted butter

1. Preheat oven to 350°F.
2. Place all ingredients into a food processor and process until finely crumbled.
3. Press mixture into a 7-inch springform pan that has been sprayed with a nonstick spray.
4. Bake for 10 minutes. Let cool for 15 minutes.

(continued on next page)

Nutritional Analysis
(per serving)

345	Calories
17.7 gm.	Fat
39.1 mg.	Cholesterol
46%	of Calories from Fat
13.3 gm.	Protein
10.19 gm.	Saturated Fat
2.07 gm.	Fiber
37.1 gm.	Carbohydrate
457.4 mg.	Sodium
92 mg.	Calcium

FILLING

8 ounces low-fat cottage cheese

8 ounces light cream cheese

½ cup egg substitute

½ cup sugar

½ teaspoon vanilla

¼ cup Dutch-style cocoa powder

¼ cup semisweet mini chocolate chips

1. Preheat oven to 350°F.
2. In a food processor with a steel blade, process the cottage cheese and cream cheese until smooth.
3. Add the egg substitute, sugar, and vanilla, and process to mix well.
4. Remove half of the mixture from the bowl. Add the cocoa powder and ¼ cup chocolate chips and blend until well mixed.
5. Pour vanilla mixture over the crust, and top with the chocolate mixture. To create the swirl, run a knife through the mixture into the springform pan.
6. Bake for 35 minutes. Let the cheesecake cool to room temperature, then refrigerate it for at least 2 hours.

Dark-and-Spicy Gingerbread

Serves 8

This gingerbread lives up to its name. It's delicious with vanilla ice cream or reduced-fat whipped topping.

3 cups Reduced Fat Bisquick

3 tablespoons instant espresso powder

3 tablespoons European-style unsweetened cocoa powder

1½ tablespoons ground ginger

1 teaspoon salt

1 teaspoon freshly ground black pepper

1½ cups packed brown sugar

1 cup egg substitute

1 tablespoon vanilla

1 cup light sour cream

½ cup molasses

1. Preheat oven to 350°F.
2. Mix the dry ingredients together.
3. Mix the liquid ingredients together.
4. Mix the liquid ingredients into the dry ingredients until smooth. Pour into 1 large or 2 small Bundt pans.
5. Bake 30 minutes for 2 cakes, 45 minutes for 1, or until a toothpick inserted into the center comes out clean.
6. Cool 10 minutes before unmolding.

Nutritional Analysis (per serving)

413	Calories
10.4 gm.	Fat
6.7 mg.	Cholesterol
23%	of Calories from Fat
12.7 gm.	Protein
5.35 gm.	Saturated Fat
12.38 gm.	Fiber
85.5 gm.	Carbohydrate
522.1 mg.	Sodium
182 mg.	Calcium

Fruit-Batter Pudding

Serves 6

This is an interesting pudding to serve for dessert or brunch.

1 pound frozen fruit (cherries, peaches, berries, apples), defrosted
1 cup egg substitute
⅓ cup sugar
1 teaspoon vanilla
¼ teaspoon almond extract
1 cup 1% milk
½ cup Reduced Fat Bisquick
3 tablespoons Kirsch or Crème de Cassis (optional)
2 tablespoons confectioners' sugar

1. Preheat oven to 350°F.
2. Spray baking dish with nonstick spray; put fruit in dish evenly.
3. Mix egg substitute with sugar, vanilla, and almond extract. Mix in milk and then Bisquick (do not overmix).
4. Pour over fruit and bake for 35 minutes, until lightly browned.
5. Remove from the oven and sprinkle Kirsch over the pudding. Serve hot or warm, sifting confectioners' sugar over the top just before serving.

Nutritional Analysis (per serving)

152	Calories
2.2 gm.	Fat
2 mg.	Cholesterol
13%	of Calories from Fat
6.7 gm.	Protein
0.62 gm.	Saturated Fat
1.06 gm.	Fiber
26.9 gm.	Carbohydrate
133 mg.	Sodium
81 mg.	Calcium

Lemon Cake

Serves 16

When you need a cake fast, this is it. You can easily create an orange cake by substituting orange extract and zest for the lemon.

4 cups Reduced Fat Bisquick

1 cup sugar plus 1 tablespoon

1 cup 1% low-fat milk

1 teaspoon lemon zest

2 teaspoons vanilla

1 teaspoon lemon extract

8 ounces egg substitute or 4 eggs

1. Preheat oven to 350°F.
2. Sift Bisquick into a large bowl, add 1 cup sugar, and mix.
3. Mix the rest of the ingredients well, reserving the 1 tablespoon of sugar. Add to the Bisquick mixture and mix until well blended.
4. Spray a large Bundt pan with nonstick spray and then sprinkle the pan with remaining sugar. Pour the batter on top of the sugar; this creates a nice top when the cake is turned out.
5. Bake for 40 minutes or use 2 smaller Bundt pans and bake for 30 minutes.

Nutritional Analysis (per serving)

115	Calories
2.2 gm.	Fat
53.2 mg.	Cholesterol
17%	of Calories from Fat
2.8 gm.	Protein
0.66 gm.	Saturated Fat
0.2 gm.	Fiber
20.6 gm.	Carbohydrate
148.4 mg.	Sodium
41 mg.	Calcium

Lemon Cheesecake

Serves 6

A lemon cheesecake for the most particular cheese cake lover! This cheesecake is best made a day ahead.

CRUST

1 cup graham cracker crumbs

¼ teaspoon cinnamon

¼ teaspoon all spice

3 tablespoons melted butter

1. Preheat oven to 350°F.
2. Place ingredients in a food processor and process until the mixture resembles fine crumbs.
3. Press the mixture into a 7-inch springform pan that has been sprayed with nonstick spray.
4. Bake for 10 minutes. Let cool for 15 minutes.

(continued on next page)

Nutritional Analysis (per serving)

301	Calories
15.1 gm.	Fat
39.1 mg.	Cholesterol
45%	of Calories from Fat
12.2 gm.	Protein
8.61 gm.	Saturated Fat
0.43 gm.	Fiber
30.3 gm.	Carbohydrate
455.8 mg.	Sodium
85 mg.	Calcium

FILLING

8 ounces low-fat cottage cheese

8 ounces light cream cheese

½ cup egg substitute

½ cup sugar

½ teaspoon vanilla

¼ teaspoon lemon oil or ½ teaspoon lemon extract

½ teaspoon lemon zest

1. Preheat oven to 350°F.
2. In a food processor with a steel blade, process the cottage cheese and cream cheese until smooth.
3. Add remaining ingredients and process to mix well.
4. Pour into a springform pan, and bake for 35 minutes.
5. Let cool to room temperature, then refrigerate at least 2 hours.

Peach Cobbler

Serves 4

Topped with vanilla yogurt, this cobbler makes an excellent breakfast. Frozen berries or any light canned fruit can replace the peaches. If you use canned fruit, use the juice from the can to replace the nectar in the recipe.

1 package (16 ounces) frozen sliced peaches (no sugar added), thawed
1 cup peach nectar
½ teaspoon cinnamon
¼ teaspoon nutmeg
2 tablespoons Cornstarch
2 cups low-fat granola cereal

1. Preheat oven to 350°F.
2. Place peaches in a soufflé dish that has been sprayed with nonstick spray.
3. Combine nectar, cinnamon, nutmeg, and cornstarch, and stir well. Pour over peaches and mix.
4. Cobble the top with the granola and bake for 30 minutes.
5. Let cool for at least 10 minutes before serving.

Nutritional Analysis (per serving)

244	Calories
5 gm.	Fat
0.0 mg.	Cholesterol
19%	of Calories from Fat
3.6 gm.	Protein
0.29 gm.	Saturated Fat
3.23 gm.	Fiber
49.7 gm.	Carbohydrate
69.3 mg.	Sodium
28 mg.	Calcium

Pear Cake

Serves 12

This is a great cake for morning coffee or afternoon tea. You can substitute peaches or apricots for the pears.

1 can (15 ounces) sliced pears in light syrup
¾ cup egg substitute or 3 eggs
½ cup light sour cream
1 teaspoon vanilla
2 teaspoons almond extract
3 cups Reduced Fat Bisquick
½ cup sugar

1. Preheat oven to 350°F.
2. Mix all liquid ingredients together.
3. Pour liquid mixture into the Bisquick and sugar, and mix well.
4. Spray two medium Bundt pans with nonstick spray. Spoon the batter into the pans.
5. Bake for 25 minutes. Cool for 10 minutes before unmolding.

Nutritional Analysis (per serving)

119	Calories
2.3 gm.	Fat
2.3 mg.	Cholesterol
17%	of Calories from Fat
2.85 gm.	Protein
0.88 gm.	Saturated Fat
1.02 gm.	Fiber
21.5 gm.	Carbohydrate
156.1 mg.	Sodium
31 mg.	Calcium

Poppy Seed–Lemon Cake

Serves 8

This Poppy Seed–Lemon Cake is a variation on cakes that I fell in love with but that were very bad for the waistline. So, for this one, I used the sponge cake!

5 egg whites
¾ cup sugar
2 eggs or 4 ounces egg substitute
2 teaspoons pure vanilla extract
1 teaspoon lemon extract
¼ cup fresh lemon juice
1 cup flour
1 teaspoon baking powder
¼ teaspoon salt
2 tablespoons poppy seeds

1. Preheat oven to 350°F.
2. In a clean bowl, beat egg whites; gradually add ¼ cup sugar until stiff peaks form. Set aside.
3. Beat the eggs with ½ cup sugar, vanilla, lemon extract, and lemon juice.
4. Sift flour, baking powder, and salt onto the top of the egg mixture. Mix well.
5. Fold 1 cup of the beaten egg whites into the flour mixture and mix to lighten. Pour batter into the remaining beaten egg whites and poppy seeds, and fold.
6. Spray a 10-inch tube pan with a nonstick spray. Pour the cake batter into the pan, and bake for 35 minutes.
7. Let cool 45 minutes before serving.

Nutritional Analysis
(per serving)

169	Calories
2.5 gm.	Fat
52.6 mg.	Cholesterol
13%	of Calories from Fat
5.7 gm.	Protein
0.55 gm.	Saturated Fat
0.8 gm.	Fiber
30.9 gm.	Carbohydrate
444.6 mg.	Sodium
76 mg.	Calcium

Pumpkin Cake

Serves 24

The perfect holiday gift, this recipe will make two large Bundt cakes or four small ones.

8 ounces egg substitute

1 cup brown sugar

½ cup sugar

1 teaspoon vanilla

1 cup skim milk

1 can (28 ounces) solid pumpkin

2 cups all-purpose flour

2 cups oatmeal

1 tablespoon baking soda

1 teaspoon baking powder

3 tablespoons pumpkin pie spice

1 cup chopped walnuts

1 cup raisins

Powdered sugar

Nutritional Analysis
(per serving)

174	Calories
4 gm.	Fat
0.3 mg.	Cholesterol
21%	of Calories from Fat
4.4 gm.	Protein
0.46 gm.	Saturated Fat
2.25 gm.	Fiber
31.6 gm.	Carbohydrate
218.3 mg.	Sodium
57 mg.	Calcium

1. Preheat oven to 350°F.
2. Mix egg substitute, brown sugar, sugar, vanilla, milk, and pumpkin in a large bowl. Set aside.
3. Mix flour, oatmeal, baking soda, baking powder, pumpkin pie spice, walnuts, and raisins.
4. Fold dry ingredients into liquid ingredients until well mixed.
5. Pour mixture into 2 large or 4 small Bundt pans sprayed with nonstick spray. Bake 45 minutes for 4 small cakes, or 60 minutes for 2 large cakes.
6. Let cool. Unmold on plates and sprinkle with powdered sugar before serving.

Pumpkin Pie with Graham Cracker Crust

Serves 8

It looks like and tastes like old-fashioned pumpkin pie. The only thing it's missing is 50 percent of the calories!

CRUST

1 cup graham cracker crumbs

3 tablespoons sugar

½ teaspoon allspice

2 tablespoons butter, melted

1. Preheat oven to 375°F.
2. Mix all ingredients together.
3. Press into a 9-inch pie pan. Bake for 8 minutes and cool.

(continued on next page)

Nutritional Analysis (per serving)

201	Calories
4.7 gm.	Fat
10 mg.	Cholesterol
21%	of Calories from Fat
4.7 gm.	Protein
2.49 gm.	Saturated Fat
1.84 gm.	Fiber
36.7 gm.	Carbohydrate
155.3 mg.	Sodium
80 mg.	Calcium

FILLING

½ cup egg substitute or 4 egg whites, slightly beaten
1½ cups solid canned pumpkin
¾ cup sugar
1 teaspoon ground cinnamon
½ teaspoon ground ginger
¼ teaspoon ground cloves
1½ cups 1% milk

1. Preheat oven to 425°F.
2. Mix all ingredients together and pour into the crust.
3. Bake 15 minutes. Reduce heat to 350°F and bake 45 minutes, or until a knife inserted into the center of the pie comes out clean. Cool before serving.

Shortcakes

Serves 8

To make the shortcakes more special, use different shaped cutters—hearts, half moons, or stars.

2¼ cups Reduced Fat Bisquick
3 tablespoons sugar
½ teaspoon vanilla extract
¼ teaspoon almond extract
⅔ cup 1% milk

1. Preheat oven to 425°F.
2. Mix ingredients together.
3. Drop by the spoonful onto ungreased cookie sheet. Or knead the dough 10 times and roll it out to about ½ inch thick. Then cut it with a cutter and place the shapes on an ungreased cookie sheet.
4. Bake 8 to 11 minutes or until golden brown.

Nutritional Analysis (per serving)

69	Calories
1 gm.	Fat
0.7 mg.	Cholesterol
14%	of Calories from Fat
1.5 gm.	Protein
0.29 gm.	Saturated Fat
0.21 gm.	Fiber
13.1 gm.	Carbohydrate
149.9 mg.	Sodium
40 mg.	Calcium

Spiced Apricots

Serves 6

These can be served hot or cold over vanilla ice milk. They are also excellent served with poultry, ham, or pork.

2 tablespoons cornstarch
2 tablespoons brown sugar
1 teaspoon pumpkin pie spice
1 tablespoon fruit-infused vinegar
1 can (28 ounces) apricot halves in light syrup, drained, reserving 1 cup liquid

1. Combine cornstarch, brown sugar, pumpkin pie spice, and vinegar in a saucepan, and bring to a boil.
2. Lower heat, add apricots, and simmer 10 minutes.

Nutritional Analysis
(per serving)

123	Calories
0.1 gm.	Fat
0.0 mg.	Cholesterol
0%	of Calories from Fat
0.8 gm.	Protein
0.01 gm.	Saturated Fat
2 gm.	Fiber
31.8 gm.	Carbohydrate
8 mg.	Sodium
21 mg.	Calcium

Tomato Soup Cake

___

Serves 8
___

The basic Tomato Soup Cake has been passed down for three generations. This time, I took a couple of liberties with the recipe to make it a bit healthier. If you like spice cake, you will love this.

2 cups sifted flour
2 teaspoons baking powder
1 teaspoon cinnamon
1 teaspoon nutmeg
½ teaspoon ground cloves
2 eggs or 4 ounces egg substitute
¾ cup sugar
1 can (10¾ ounces) tomato soup
½ cup chopped walnuts

1. Preheat oven to 350°F.
2. Sift flour, baking powder, cinnamon, nutmeg, and cloves together and set aside.
3. Combine eggs, sugar, and tomato soup. Gradually add flour mixture to liquid mixture. Stir in walnuts.
4. Pour into a Bundt pan that has been sprayed with nonstick spray. Bake for 25 to 30 minutes.

Nutritional Analysis (per serving)

260	Calories
6.9 gm.	Fat
52.6 mg.	Cholesterol
24%	of Calories from Fat
6 gm.	Protein
1.02 gm.	Saturated Fat
1.31 gm.	Fiber
44.6 gm.	Carbohydrate
255.4 mg.	Sodium
87 mg.	Calcium

Index

A

Acorn squash, 81
Allspice, 11, 23, 35, 36
Almond extract, 39
Almonds, 179
Altitude, 2
Anise, 12, 18, 39
Aniseed, 18
Appetizers
    about, 42
    Artichokes with Shrimp, 54
    Black Bean Spread, 88
    Cheddar Cheese Spread, 43
    Cheddar Olive Spread, 44
    Chicken Drumettes, 55
    Chicken Liver Pâté, 51
    Country Pâté without Liver, 52
    Crab Cakes, 56
    Eggplant Caviar, 48
    Feta Cheese Spread, 45
    Goat-Cheese Spread, 46, 57
    Guacamole, 49
    Hummus, 50
    Phyllo Goat Cheese Squares, 57
    Spinach and Chicken Pâté, 53
    Vegetable Cream Cheese Spread, 47
Apple-Walnut Cake, 229
Apples
    Apple-Walnut Cake, 229
    Fruit Salad with Dressing, 67
    Fruit-Batter Pudding, 237
    Waldorf Salad, 76
Apricot Nut Bread, 211
Apricots
    Apricot Nut Bread, 211
    Fruit Salad with Dressing, 67
    Spiced Apricots, 248
Artichokes
    Artichokes with Shrimp, 54
    Fettuccine with Artichokes and Tomatoes, 98
    Shrimp-Stuffed Artichokes, 123
    steamed, 175
Artichokes with Shrimp, 54
Asparagus, 35, 151, 182
Author contact information, 2
Avocados, 49

B

Baked
    Chicken, 132
    Halibut, 115
    Oatmeal, 203
    Sole with Shrimp and Crabmeat, 116
Balsamic Vinegar, 75
Barbeque
    Chicken with Vinegar-Basil Marinade, 133
    Swordfish, 117
Barley, 82
Basic Herb Blend
    Artichokes, 175
    Baked Halibut, 115
    Brussels sprouts, 176
    Corn Soup, 79
    Crab Cakes, 56
    making, 40
Basic Marinara Sauce, 95
Basic Sponge Cake, 230
Basil
    about, 12–13
    BBQ Chicken with Vinegar-Basil Marinade, 133
    Chicken Herb Blend, 40
    Herb Salt, 40
    Italian Herb Blend, 40
    Orzo with Parmesan and Basil, 195
    Tomato-Basil Sauce with Mostaccioli, 109
    usage chart, 35–38
    Vegetable Herb Blend, 40
    Vermicelli with Basil and Pine Nuts, 110
Bay leaf
    about, 13–14
    usage chart, 35–38
BBQ
    Chicken with Vinegar-Basil Marinade, 133
    Swordfish, 117
Beans
    about, 86
    Black Bean Spread, 88
    Black-and-White Bean Salad, 87
    Green beans, 69, 174, 178, 179, 180
    Green Beans Almondine, 179
    Green Beans Vinaigrette, 69, 180
    Lentil Stew, 89
    Navy Bean and Basil Salad, 90

251

Beans (*Continued*)
    Navy Bean Soup, 91
    Red Beans and Rice Salad, 92
    Turkey Chili, 153
Beef
    about, 158
    Chateubriand, 159
    Flank Steak a la India, 160
    Ground Beef with Rice, 161
    Meat Sauce, 102
    Meatloaf, 165
    seasoning chart, 38
    Simple Beef Stew, 172
    Soy-Ginger Marinade, 131
Beer, 212
Beets, 35
Berries
    Fruit Salad with Dressing, 67
    Fruit-Batter Pudding, 237
    harvesting, 6
Bibb Lettuce with Radishes, 63
Biscuits, 222, 247
Black Bean Spread, 88
Black beans
    Black Bean Spread, 88
    Black-and-White Bean Salad, 87
Black Forest Chocolate Cake, 231
Black pepper
    Greek Herb Blend, 40
    usage chart, 35–38
Black-and-White Bean Salad, 87
Blends, herb. *See also* individual herb blends
    Basic
        Artichokes, 175
        Baked Halibut, 115
        Brussels sprouts, 176
        Corn Soup, 79
        Crab Cakes, 56
    Bouquet garni, 28
    Chicken, 40
    Dip, 40
    Fines herbes, 17
    Greek, 40
    Herb Salt, 40
    Italian
        Baked Chicken, 132
        Basic Marinara Sauce, 95
        Green Beans Vinaigrette, 69, 180
        Ground Beef with Rice, 161
        Meat Sauce, 102
        Meatloaf, 165
        Pot-Roasted Pasta, 105
        Potato Frittata, 206
        Rolled Chicken with Asparagus, 151
        Shrimp Scampi Provincial with Linguine, 107
        Simple Beef Stew, 172
        Turkey Sausage with Peppers and Onions, 156
        Vermicelli with Cream Cheese and Herbs, 111
    making, 39–40
    Mexican, 40
    Vegetable
        Cheddar-Beer Bread, 212
        Chicken Stew, 143
        Cucumbers and Tomatoes in Lettuce Cups, 65
        Fried Rice, 197
        Goat-Cheese Spread, 46
        Herbed Spoon Bread, 218
        Macaroni and Cheese, 101
        Orange and Onion Chicken, 148
        Oven-Fried Chicken, 150
        Savory Biscuits, 222
        Tricolor Peppers, 188
Bouquet garni, 28
Brandy extract, 39
Breads
    about, 210
    Apricot Nut Bread, 211
    Cheddar-Beer Bread, 212
    Cheese Monkey Bread, 213
    Chocolate Chip and Walnut Scones, 214
    Cinnamon Nut Bread, 215
    Cinnamon Rolls, 216
    Green Chili Cornbread, 217
    Herbed Spoon Bread, 218
    Nut Bread, 219
    Poppy Seed Rolls, 220
    Pumpkin Muffins, 221
    Savory Biscuits, 222
    Scones, 223
    Waffles, 208
Breakfast
    Baked Oatmeal, 203
    Crustless Quiche, 204
    Peanut Butter Breakfast Bars, 205
    Potato Frittata, 206
    Veggie Omelet, 207
    Waffles, 208
Broccoli, 35
Brownies, 232
Brussels Sprouts, 35, 144, 176
Butternut squash, 81

C

Cabbage
    Cabbage-Carrot Coleslaw, 64
    herb chart, 35
    Spinach and Red Cabbage Salad with Honey-Mustard Dressing, 73
Cabbage-Carrot Coleslaw, 64

Index

Cacciatore
    Chicken, 136
    Halibut, 120
Cajun Chicken, 134
Cakes
    Apple-Walnut Cake, 229
    Basic Sponge Cake, 230
    Black Forest Chocolate Cake, 231
    Brownies, 232
    Chocolate Swirl Cheesecake, 234–235
    Chocolate-Chip Sponge Cake, 233
    Dark-and-Spicy Gingerbread, 236
    Lemon Cake, 238
    Lemon Cheesecake, 239–240
    Pear Cake, 242
    Poppy Seed-Lemon Cake, 243
    Pumpkin Cake, 244
    Shortcakes, 247
    Tomato Soup Cake, 249
Calories, salad ingredients, 60–62
Canadian bacon, 91
Cantaloupe, 67
Capers, 14, 146
Caraway seed, 15, 35
Cardomom, 15–16, 23, 36
Carrots
    about, 174
    Cabbage-Carrot Coleslaw, 64
    Chervil, 17
    Dilled Carrots, 177
    herb chart, 36
    Julienne Carrots and Zucchini, 181
    New Potatoes and Carrots, 184
    storing, 174
Cassia, 21
Cauliflower, 35
Cayenne, 16, 40
Celery
    salt, 37, 38
    seed, 17, 36, 40
    storing, 174
Chart, herb and spice, 35–38
Chateubriand, 159
Cheddar cheese
    Cheddar Cheese Spread, 43
    Cheddar Olive Spread, 44
    Cheddar-Beer Bread, 212
    Cheese Monkey Bread, 213
    Macaroni and Cheese, 101
Cheddar Cheese Spread, 43
Cheddar Olive Spread, 44
Cheddar-Beer Bread, 212
Cheese
    Cheddar Cheese Spread, 43
    Cheddar Olive Spread, 44

Cheddar-Beer Bread, 212
Cheese Monkey Bread, 213
Chocolate Swirl Cheesecake, 234–235
Cinnamon Rolls, 216
Colorful Pasta Peppers, 96
Crustless Quiche, 204
Feta Cheese Spread, 45
Fettuccine Alfredo, 97
Goat-Cheese Spread, 46, 57
Greek Salad, 68
Greek-Style Orzo, 194
Greek-Style Shrimp, 119
Guacamole, 49
Lemon Cheesecake, 239–240
Macaroni and Cheese, 101
Orzo with Parmesan and Basil, 195
Phyllo Goat Cheese Squares, 57
Spinach and Feta, 72
Tomato Basil with Fresh Mozzarella and Balsamic
    Vinegar, 75
Turkey Sausage with Peppers and Onions, 156
Vegetable Cream Cheese Spread, 47
Vermicelli with Cream Cheese and Herbs, 111
Yogurt, 128, 227
Cheese Monkey Bread, 213
Cheesecake, 234–235, 239–240
Cherries
    Black Forest Chocolate Cake, 231
    Fruit-Batter Pudding, 237
Chervil, 17–18, 40
Chicken
    about, 128–129
    Baked Chicken, 132
    BBQ Chicken with Vinegar-Basil Marinade, 133
    Cajun Chicken, 134
    Chicken Breast Supreme, 135
    Chicken Cacciatore, 136
    Chicken Cakes with Tomato and Sweet Pepper
        Sauce, 137–138
    Chicken Chasseur, 139
    Chicken Drumettes, 55
    Chicken in Lettuce Cups, 140
    Chicken Liver Pâté, 51
    Chicken Polynesian, 141
    Chicken Polynesian II, 142
    Chicken Stew, 143
    Chicken Strips with Brussels Sprouts, 144
    Chicken Thighs with an Indian Flair, 145
    Chicken with Lemon and Capers, 146
    Herb Blend, 40
    Marinade, 130
    Mustard Chicken, 147
    Orange and Onion Chicken, 148
    Orange-Rosemary Cornish Game Hens, 149
    Oven-Fried Chicken, 150

Index

Chicken (*Continued*)
   Rolled Chicken with Asparagus, 151
   seasoning chart, 36
   Soy-Ginger Marinade, 131
   Spinach and Chicken Pâté, 53
Chicken Breast Supreme, 135
Chicken Cacciatore, 136
Chicken Cakes with Tomato and Sweet Pepper Sauce, 137–138
Chicken Chasseur, 139
Chicken Drumettes, 55
Chicken in Lettuce Cups, 140
Chicken Liver Pâté, 51
Chicken Polynesian, 141
Chicken Polynesian II, 142
Chicken Stew, 143
Chicken Strips with Brussels Sprouts, 144
Chicken Thighs with an Indian Flair, 145
Chicken with Lemon and Capers, 146
Chickpeas, 50
Chiles, 16, 217
Chili, 153
Chili powder, 36, 38
China
   Eight-Spice Powder, 18
   Five-Spice Powder, 18
   flavorings, 4
Chives, 18–19, 35–37, 40
Chocolate
   Black Forest Chocolate Cake, 231
   Brownies, 232
   Chocolate Chip and Walnut Scones, 214
   Chocolate Chip Sponge Cake, 233
   Chocolate Swirl Cheesecake, 234–235
Chocolate Chip and Walnut Scones, 214
Chocolate Chip Sponge Cake, 233
Chocolate Swirl Cheesecake, 234–235
Cilantro, 19–21
Cinnamon, 18, 20–21, 36, 215, 216
Cinnamon Nut Bread, 215
Cinnamon Rolls, 216
Citrus
   Chicken with Lemon and Capers, 146
   Lemon Cake, 238
   Lemon Cheesecake, 239–240
   Lemon Rice, 199
   Lemon zest, 29, 152
   Lemon-Herbed Asparagus, 182
   Orange and Onion Chicken, 148
   Orange zest, 33, 148
   Orange-Rosemary Cornish Game Hens, 149
   Poppy Seed-Lemon Cake, 243
   Turkey Breast Marinated with Lemon and Herbs, 152
Clams
   Linguine with Clam Sauce, 99
   Linguine with Herbed Clam Sauce, 100
   usage chart, 37

Cloves, 18, 21–22, 35
Cobbler, Peach, 241
Cold soup, 80
Coleslaw, 64
Colorful Pasta Peppers, 96
Contact information, author, 2
Cookies
   Brownies, 232
   Shortcakes, 247
Coriander, 19, 23
Corn
   Corn Soup, 79
   Fried Rice, 197
   Green Chili Cornbread, 217
   usage chart, 36
Corn Soup, 79
Cornish Game Hens, 36, 149
Cottage Cheese
   Cheddar Cheese Spread, 43
   Cheddar Olive Spread, 44
   Chocolate Swirl Cheesecake, 234–235
   Cinnamon Rolls, 216
   Crustless Quiche, 204
   Feta Cheese Spread, 45
   Guacamole, 49
Country Pâté without Liver, 52
Crab
   Baked Sole with Shrimp and Crabmeat, 116
   boil, 37
   Crab Cakes, 56
   Pasta Salad with Crab and Snow Peas, 104
   usage chart, 37
Crab Cakes, 56
Crayfish, 37
Cream Cheese, 46, 47, 51, 111, 234–235
Crust, Graham Cracker, 234, 239, 245
Crustless Quiche, 204
Cucumber and Tomatoes in Lettuce Cups, 65
Cucumbers
   Cucumber and Tomatoes in Lettuce Cups, 65
   Gazpacho, 80
   usage chart, 36
   Yogurt-Cucumber Soup, 84
Cumin seed, 22–23, 38, 40
Currants, 203, 223
Curry, 4, 23, 38

D

Dark-and-Spicy Gingerbread, 236
Date Risotto, 196
Dates
   Date Risotto, 196
   Pumpkin Muffins, 221
Desserts
   about, 226
   Apple-Walnut Cake, 229

Index

Basic Sponge Cake, 230
Black Forest Chocolate Cake, 231
Brownies, 232
Chocolate Swirl Cheesecake, 234–235
Chocolate-Chip Sponge Cake, 233
Dark-and-Spicy Gingerbread, 236
Fruit-Batter Pudding, 237
Lemon Cake, 238
Lemon Cheesecake, 239–240
Peach Cobbler, 241
Pear Cake, 242
Poppy Seed-Lemon Cake, 243
Pumpkin Cake, 244
Pumpkin Pie with Graham Cracker Crust, 245–246
Shortcakes, 247
Spiced Apricots, 248
Tomato Soup Cake, 249
Yogurt Cheese, 227
Yogurt Cream, 228
Dill, 23–24, 35–38, 40, 66, 177
Dilled Carrots, 177
Dilled Shrimp Salad, 66
Dip Herb Blend, 40
Dips. *See* Spreads
Dressing, salad, 67, 69, 73
Drumettes, Chicken, 55
Dry-Curd Cottage Cheese, 43, 44, 45
Drying herbs, 7–8. *See also* individual herbs
Duck, 36

E

Eggplant
Eggplant Caviar, 48
Moussaka, 166–167
Oven-Fried Eggplant, 185
usage chart, 35
Eggplant Caviar, 48
Eggs
Crustless Quiche, 204
Herbed Spoon Bread, 218
Potato Frittata, 206
Veggie Omelet, 207
Eight Spice Powder, Chinese, 18
Ethnicity, herbs, 4
Europe, flavorings, 4
Extracts, 38–39

F

Fennel, 18, 24–25, 37
Feta cheese
Feta Cheese Spread, 45
Greek Salad, 68
Greek-Style Orzo, 194

Greek-Style Shrimp, 119
Spinach and Feta, 72
Fettuccine Alfredo, 97
Fettuccine with Artichokes and Tomatoes, 98
Fines herbes, 17
Fish
about, 114
Baked Halibut, 115
Baked Sole with Shrimp and Crabmeat, 116
BBQ Swordfish, 117
Cajun, 134
Ginger-Orange Halibut, 118
Halibut Cacciatore, 120
Poached Halibut and Peppers, 121
Sole Veronique, 124
usage chart, 37
Five-Spice Powder, Chinese, 18
Flank Steak a la India, 160
Flowers, harvesting, 6
Fowl. *See* Poultry
Freezing herbs, 8. *See also* individual herbs
French Fries, 192
Freshness, herbs, 10
Fried Chicken, 150
Fried Eggplant, 185
Fried Rice, 197
Fried Zucchini and Yellow Squash, 187
Frittata, 206
Fruit
Apple-Walnut Cake, 229
Apricot Nut Bread, 211
Baked Oatmeal, 203
Black Forest Chocolate Cake, 231
Chicken Polynesian, 141
Chicken Polynesian II, 142
Chicken with Lemon and Capers, 146
Date Risotto, 196
Fruit Salad with Dressing, 67
Fruit-Batter Pudding, 237
harvesting, 6
Indian Raisin Rice, 198
Lemon Cake, 238
Lemon Cheesecake, 239–240
Lemon Rice, 199
Lemon zest, 29, 152
Lemon-Herbed Asparagus, 182
Orange and Onion Chicken, 148
Orange zest, 33, 148
Orange-Rosemary Cornish Game Hens, 149
Peach Cobbler, 241
Pear Cake, 242
Poppy Seed-Lemon Cake, 243
Pumpkin Cake, 244
Pumpkin Muffins, 221
Romaine, Pears, and Walnuts, 71
Scones, 223
Spiced Apricots, 248

Index

Fruit (*Continued*)
    Spinach, Raspberry, and Walnut Salad, 74
    Turkey Breast Marinated with Lemon and Herbs, 152
    Waldorf Salad, 76
Fruit Salad with Dressing, 67
Fruit-Batter Pudding, 237

G

Game, 38
Gamebirds, 36, 37, 149
Garbanzo beans. *See* Chickpeas
Gardens, herb, 4–5. *See also* individual herbs
Garlic
    about, 25–26
    powder, 35–38, 40
    salt, 35–38
Gazpacho, 80
Ginger
    about, 27–28
    Cardomom, 15
    Dark-and-Spicy Gingerbread, 236
    Eight-spice powder, 18
    Ginger-Orange Halibut, 118
    Mexican Herb blend, 40
    Soy-Ginger Marinade, 131
    usage chart, 36, 38
Ginger-Orange Halibut, 118
Gingerbread, 236
Goat-Cheese Spread, 46, 57
Golden Squash Soup, 81
Graham Cracker Crust, 234, 239, 245
Grapes, 124
Gravy, 155
Greek
    flavorings, 4
    Herb Blend, 40
    Orzo, 194
    Shrimp, 119
Green beans, 36, 69, 174, 178, 179, 180
Green Beans Almondine, 179
Green Beans Vinaigrette, 69, 180
Green Chili Cornbread, 217
Ground Beef with Rice, 161
Growing herbs, 4–5. *See also* individual herbs
Guacamole, 49

H

Halibut, 115, 118, 120, 121
Halibut Cacciatore, 120
Harvesting herbs, 6
Hens, Cornish, 36, 149
Herb Salt, 40
Herb-Crusted Lamb, 162

Herbed Spoon Bread, 218
Herbs. *See also* Spices; individual herbs
    Allspice, 35, 36
    Anise, 12, 39
    Basic Herb blend, 40
    Basil, 12–13, 35–38, 40
    Bay leaf, 13–14, 35–38
    Black pepper, 35–38, 40
    Blends, making, 39–40 (*See also* individual herb blends)
    Bouquet garni, 28
    Capers, 14
    Caraway seed, 15, 35
    Cardomom, 36
    Cayenne, 40
    Celery salt, 36–38
    Celery seed, 17, 36, 40
    Chart, 35–38
    Chervil, 17–18, 40
    Chicken Herb blend, 40
    Chili powder, 36, 38
    Chives, 18–19, 35–37, 40
    Cilantro, 19–20
    Cinnamon, 36
    Cloves, 35
    Crab and shrimp boil, 37
    Cumin seed, 22–23, 38, 40
    Curry, 38
    definition, 3–4
    Dill, 23–24, 35–38, 40
    Dip Herb blend, 40
    drying, 7–8
    ethnicity, 4
    Extracts, 38–39
    Fennel, 18, 24–25, 37
    freezing, 8
    freshness, 10
    gardens, 4–5 (*See also* individual herbs)
    Garlic, 25–26
    Garlic powder, 35–38, 40
    Garlic salt, 35–38
    Ginger, 36, 38, 40
    Greek Herb blend, 40
    harvesting, 6
    history, 3, 4
    infused oils, 9
    Italian Herb blend, 40
    Italian seasoning, 35–37
    Juniper berries, 38
    Leeks, 28
    Lemon and pepper seasoning salt, 35–37
    Lemon peel, 40
    Mace, 36
    Marjoram, 29–30, 33, 35–38, 40
    Meat tenderizer, 38
    Mexican Herb blend, 40

microwaving, 8
Mint, 30–31, 33, 36
Mustard, 31–32, 35–38, 40
Nutmeg, 35–36
Onion, 36–38, 40
Oregano, 30, 33, 35–38, 40
Paprika, 35–37
Parsley, 17, 34–38, 40
Pepper, 35–38, 40
Peppercorns, 37
Poultry seasoning, 36
preserving, 6–9
Pumpkin Pie spice, 36
Red pepper, 35–38, 40
Rosemary, 35–38, 40
Saffron, 36
Sage, 36
Salt, 35, 36, 38, 40
Salt Herb blend, 40
Seafood seasoning, 37
Seasoned salt, 35, 36, 38
Sesame seed, 36, 38
Shrimp boil, 37
storing, 6–10
Tarragon, 36, 37
Thyme, 35–38, 40
usage chart, 35–38
Vegetable Herb blend, 40
White pepper, 35, 37
History, herbs and spices, 3, 4
Honey-Mustard Dressing, 73
Hoop Cheese, 43, 44, 45
Hummus, 50

I

India flavorings, 4
Indian Raisin Rice, 198
Indonesia, flavorings, 4
Infused oils, 9
Italian Herb Blend
    Baked Chicken, 132
    Basic Marinara Sauce, 95
    Green Beans Vinaigrette, 69, 180
    Ground Beef with Rice, 161
    making, 40
    Meat Sauce, 102
    Meatloaf, 165
    Potato Frittata, 206
    Rolled Chicken with Asparagus, 151
    Simple Beef Stew, 172
    Turkey Sausage with Peppers and Onions, 156
    Vermicelli with Cream Cheese and Herbs, 111
Italian seasoning, 35–37
Italy, flavorings, 4

J

Jambalaya, Pork, 168
Julienne Carrots and Zucchini, 181
Juniper berries, 38

K

Kebabs, lamb, 163

L

Lamb
    about, 158
    Herb-Crusted Lamb, 162
    Lamb Kebabs, 163
    Lamb Loin, 164
    Moussaka, 166–167
    usage chart, 38
Lamb Kebabs, 163
Lamb Loin, 164
Laurel, 13–14, 21
Leaves, harvesting, 6
Leeks, 28
Lemon
    Chicken with Lemon and Capers, 146
    extract, 39
    Lemon Cake, 238
    Lemon Cheesecake, 239–240
    Lemon Rice, 199
    Lemon-Herbed Asparagus, 182
    mint, 30
    peel, 40
    pepper seasoning salt, 35–37
    Poppy Seed-Lemon Cake, 243
    Turkey Breast Marinated with Lemon and Herbs, 152
    zest, 29, 152
Lemon Cake, 238
Lemon Cheesecake, 239–240
Lemon Rice, 199
Lemon-Herbed Asparagus, 182
Lentil Stew, 89
Lettuce
    Bibb Lettuce with Radishes, 63
    Black-and-White Bean Salad, 87
    Chicken in Lettuce Cups, 140
    Cucumber and Tomatoes in Lettuce Cups, 65
    Dilled Shrimp Salad, 66
    Fruit Salad with Dressing, 67
    Greek Salad, 68
    Green Beans Vinaigrette, 69, 180
    Navy Bean and Basil Salad, 90
    Rice Salad Oriental, 70
    Romaine, Pears, and Walnuts, 71

Index

Lettuce (*Continued*)
    Shrimp-Stuffed Artichokes, 123
    Tomato Basil with Fresh Mozzarella and Balsamic Vinegar, 75
Licorice, 18
Lilies, 28
Linguine
    Clam Sauce, 99
    Herbed Clam Sauce, 100
Liver, chicken, 51
Lobster, 37

M

Macaroni and Cheese, 101
Mace, 29, 32, 36
Mandarin oranges, 54, 123, 142
Marinades
    Basic, 130
    BBQ Chicken with Vinegar-Basil Marinade, 133
    Soy-Ginger, 131
    Turkey Breast Marinated with Lemon and Herbs, 152
Marinara Sauce, 95
Marjoram, 29–30, 33, 35–38, 40
Mashed Potatoes, 191
Meat Sauce, 102
Meat tenderizer, 38
Meatloaf, 165
Meats. *See also* Poultry
    about, 158
    Cajun, 134
    Chateaubriand, 159
    Flank Steak a la India, 160
    Ground Beef with Rice, 161
    Herb-Crusted Lamb, 162
    Lamb Kebabs, 163
    Lamb Loin, 164
    Meat Sauce, 102
    Meatloaf, 165
    Moussaka, 166–167
    Pork Jambalaya, 168
    Pork with Tomato-Rosemary Sauce, 169
    Prosciutto and Tomato Pasta, 106
    Roast Pork Tenderloin with Sage, 170
    Roast Pork with Brown Rice, 171
    Simple Beef Stew, 172
    Soy-Ginger Marinade, 131
    usage chart, 38
Mexico
    flavorings, 4
    Mexican Herb Blend, 40
Microwaving herbs, 8
Mint, 12–13, 30–31, 33, 36
Monkey Bread, 213
Monterey Jack cheese, 204

Moussaka, 166–167
Mozzarella cheese
    Cheddar-Beer Bread, 212
    Colorful Pasta Peppers, 96
    Crustless Quiche, 204
    Tomato Basil with Fresh Mozzarella and Balsamic Vinegar, 75
    Turkey Sausage with Peppers and Onions, 156
Muffins, 221
Mushroom and Barley Soup, 82
Mushrooms
    Baked Halibut, 115
    Chicken Breast Supreme, 135
    Chicken Cacciatore, 136
    Chicken Chasseur, 139
    Chicken in Lettuce Cups, 140
    Halibut Cacciatore, 120
    Lamb Kebabs, 163
    Mushroom and Barley Soup, 82
    Mushrooms Paprika, 183
    Pot-Roasted Pasta, 105
    Scallops in White Wine, 122
    Snow Peas with Mushrooms, 186
    Veggie Omelet, 207
Mushrooms Paprika, 183
Mussels, 37
Mustard, 23, 31–32, 35–38, 40, 73
Mustard Chicken, 147
Mutton. *See* Lamb
Myrtle, 21

N

Navy Bean and Basil Salad, 90
Navy Bean Soup, 91
Navy beans
    Black-and-White Bean Salad, 87
    Navy Bean and Basil Salad, 90
    Navy Bean Soup, 91
New Potatoes and Carrots, 184
Nightshades, 16
Nut Bread, 219
Nutmeg, 29, 32, 35, 36
Nuts
    Apple-Walnut Cake, 229
    Apricot Nut Bread, 211
    Chicken Polynesian II, 142
    Chocolate Chip and Walnut Scones, 214
    Cinnamon Nut Bread, 215
    Cinnamon Rolls, 216
    Date Risotto, 196
    Green Beans Almondine, 179
    Nut Bread, 219
    Peanut Butter Breakfast Bars, 205
    Pumpkin Cake, 244

Pumpkin Muffins, 221
Romaine, Pears, and Walnuts, 71
Spinach and Feta, 72
Spinach, Raspberry, and Walnut Salad, 74
Tomato Soup Cake, 249
Vermicelli with Basil and Pine Nuts, 110
Waldorf Salad, 76
Yogurt-Cucumber Soup, 84

O

Oats, 203, 205, 219, 244
Oils
    extracts, 38–39
    infused, 9
Olives
    Basic Marinara Sauce, 95
    Cheddar Olive Spread, 44
    Navy Bean and Basil Salad, 90
    Pot-Roasted Pasta, 105
Omelet, 207
Onions
    Chives, 18
    Garlic, 25–26
    Leeks, 28
    Orange and Onion Chicken, 148
    powder, 36–38, 40
    Turkey Sausage with Peppers and Onions, 156
Orange
    Artichokes with Shrimp, 54
    Chicken Polynesian II, 142
    extract, 39
    Ginger-Orange Halibut, 118
    Orange and Onion Chicken, 148
    Orange-Rosemary Cornish Game Hens, 149
    Shrimp-Stuffed Artichokes, 123
    zest, 33, 148
Orange and Onion Chicken, 148
Orange-Rosemary Cornish Game Hens, 149
Oregano, 30, 33, 35–38, 40
Orzo, 194, 195
Oven French Fries, 192
Oven-Fried Chicken, 150
Oven-Fried Eggplant, 185
Oysters, 37

P

Paprika, 33–34, 35–37, 40, 183
Parmesan cheese
    Fettuccine Alfredo, 97
    Orzo with Parmesan and Basil, 195
Parsley, 17, 22, 34, 35–38, 40
Pasta
    about, 94, 190

Basic Marinara Sauce, 95
Colorful Pasta Peppers, 96
Fettuccine Alfredo, 97
Fettuccine with Artichokes and Tomatoes, 98
Greek-Style Orzo, 194
Linguine with Clam Sauce, 99
Linguine with Herbed Clam Sauce, 100
Macaroni and Cheese, 101
Meat Sauce, 102
Orzo with Parmesan and Basil, 195
Pasta Salad with Crab and Snow Peas, 104
Pasta with Shrimp, Zucchini, and Tomatoes, 103
Pot-Roasted Pasta, 105
Prosciutto and Tomato Pasta, 106
Shrimp Scampi Provincial with Linguine, 107
Sun-Dried Tomatoes, Rosemary, and Thyme Pasta, 108
Tomato-Basil Sauce with Mostaccioli, 109
Vermicelli with Basil and Pine Nuts, 110
Vermicelli with Cream Cheese and Herbs, 111
Pasta Salad with Crab and Snow Peas, 104
Pasta with Shrimp, Zucchini, and Tomatoes, 103
Pâté
    Chicken Liver, 51
    Country Pâté without Liver, 52
    Spinach and Chicken Pâté, 53
Peach Cobbler, 241
Peaches
    Fruit-Batter Pudding, 237
    Peach Cobbler, 241
Pear Cake, 242
Pears, 242
Peas, Snow, 36, 104, 186
Pepper, 16, 35–38, 40
Peppercorns, 37
Peppermint, 30
Peppers
    about, 174
    Chicken Cakes with Tomato and Sweet Pepper Sauce, 137–138
    Colorful Pasta Peppers, 96
    Poached Halibut and Peppers, 121
    Szechwan, 18
    Tricolor Peppers, 57, 188
    Turkey Sausage with Peppers and Onions, 156
Phyllo Goat Cheese Squares, 57
Picking herbs, 6
Pie, Pumpkin with Graham Cracker Crust, 245–246
Pine Nuts, 110
Pineapple
    Chicken Polynesian, 141
    Chicken Polynesian II, 142
Poached Halibut and Peppers, 121
Poppy Seed
    Lemon Cake, 243
    Rolls, 220

Index

Pork
    about, 158
    Country Pâté without Liver, 52
    Pork Jambalaya, 168
    Pork with Tomato-Rosemary Sauce, 169
    Roast Pork Tenderloin with Sage, 170
    Roast Pork with Brown Rice, 171
    seasoning chart, 38
Pork Jambalaya, 168
Pork with Tomato-Rosemary Sauce, 169
Pot-Roasted Pasta, 105
Potato Frittata, 206
Potato Pie, 193
Potatoes
    about, 190
    Mashed Potatoes, 191
    New Potatoes and Carrots, 184
    Oven French Fries, 192
    Potato Frittata, 206
    Potato Pie, 193
    Sweet, 36
    usage chart, 35–36
Poultry
    about, 128–129
    Baked Chicken, 132
    BBQ Chicken with Vinegar-Basil Marinade, 133
    Cajun Chicken, 134
    Chicken Breast Supreme, 135
    Chicken Cacciatore, 136
    Chicken Cakes with Tomato and Sweet Pepper Sauce, 137–138
    Chicken Chasseur, 139
    Chicken Drumettes, 55
    Chicken Herb Blend, 40
    Chicken in Lettuce Cups, 140
    Chicken Liver Pâté, 51
    Chicken Polynesian, 141
    Chicken Polynesian II, 142
    Chicken Stew, 143
    Chicken Strips with Brussels Sprouts, 144
    Chicken Thighs with an Indian Flair, 145
    Chicken with Lemon and Capers, 146
    Country Pâté without Liver, 52
    Marinade, 130
    Mustard Chicken, 147
    Orange and Onion Chicken, 148
    Orange-Rosemary Cornish Game Hens, 149
    Oven-Fried Chicken, 150
    Rolled Chicken with Asparagus, 151
    seasoning, 36
    seasoning chart, 36–37
    Soy-Ginger Marinade, 131
    Spinach and Chicken Pâté, 53
    Turkey Breast Marinated with Lemon and Herbs, 152
    Turkey Chili, 153
    Turkey Sausage, 154, 155, 156
Prawns. *See* Shrimp
Preserving herbs, 6–9. *See also* individual herbs
Prosciutto and Tomato Pasta, 106
Pruning herbs, 6
Pudding, Fruit-Batter, 237
Pumpkin
    Brownies, 232
    Pumpkin Cake, 244
    Pumpkin Muffins, 221
    Pumpkin Pie with Graham Cracker Crust, 245–246
Pumpkin Cake, 244
Pumpkin Muffins, 221
Pumpkin Pie spice, 36, 244, 248
Pumpkin Pie with Graham Cracker Crust, 245–246

Q

Quiche, 204
Quick breads. *See* Breads

R

Radishes
    Bibb Lettuce with Radishes, 63
    Rice Salad Oriental, 70
Raisins, 198, 203, 244
Raspberries, 74
Red Beans and Rice Salad, 92
Red pepper, 35–38, 40
Rice
    about, 190
    Date Risotto, 196
    Fried Rice, 197
    Ground Beef with Rice, 161
    Indian Raisin Rice, 198
    Lemon Rice, 199
    Pork Jambalaya, 168
    Red Beans and Rice Salad, 92
    Roast Pork with Brown Rice, 171
    Wild-Brown Rice, 200
Rice Salad Oriental, 70
Risotto, 196
Roast Pork Tenderloin with Sage, 170
Roast Pork with Brown Rice, 171
Roasting garlic, 26
Rolled Chicken with Asparagus, 151
Rolls
    Cinnamon, 216
    Poppy Seed Rolls, 220
Romaine, Pears, and Walnuts, 71
Roots, harvesting, 6
Rosemary
    about, 35–38, 40

Index

Orange-Rosemary Cornish Game Hens, 149
Pork with Tomato-Rosemary Sauce, 169
Sun-Dried Tomatoes, Rosemary, and Thyme Pasta, 108
Round-leafed mint, 30
Rum extract, 39

S

Saffron, 23, 36
Sage
    about, 36
    Roast Pork Tenderloin with Sage, 170
Salads
    about, 60–62
    basic ingredients, 60–62
    Bibb Lettuce with Radishes, 63
    Black-and-White Bean Salad, 87
    Cabbage-Carrot Coleslaw, 64
    calories, 60–62
    Cucumber and Tomatoes in Lettuce Cups, 65
    Dilled Shrimp Salad, 66
    dressing, 73
    Fruit Salad with Dressing, 67
    Greek Salad, 68
    Green Beans Vinaigrette, 69, 180
    Navy Bean and Basil Salad, 90
    Pasta Salad with Crab and Snow Peas, 104
    Red Beans and Rice Salad, 92
    Rice Salad Oriental, 70
    Romaine, Pears, and Walnuts, 71
    Shrimp-Stuffed Artichokes, 123
    Spinach and Feta, 72
    Spinach and Red Cabbage Salad with Honey-Mustard Dressing, 73
    Spinach, Raspberry, and Walnut Salad, 74
    Tomato Basil with Fresh Mozzarella and Balsamic Vinegar, 75
    Waldorf Salad, 76
Salt, 35, 36, 38, 40
Salt, Herb, 40
Sauces
    Basic Marinara Sauce, 95
    Chicken Cakes with Tomato and Sweet Pepper Sauce, 137–138
    Gravy, 155
    Marinade, 130
    Meat Sauce, 102
    Pork with Tomato-Rosemary Sauce, 169
    Tomato-Basil Sauce with Mostaccioli, 109
Sausage, 154, 155, 156
Savory Biscuits, 222
Scallops, 37, 122
Scallops in White Wine, 122
Scones, 214, 223

Seafood
    about, 114
    Artichokes with Shrimp, 54
    Baked Halibut, 115
    Baked Sole with Shrimp and Crabmeat, 116
    BBQ Swordfish, 117
    Cajun, 134
    Crab Cakes, 56
    Ginger-Orange Halibut, 118
    Greek-Style Shrimp, 119
    Halibut Cacciatore, 120
    Linguine with Clam Sauce, 99
    Linguine with Herbed Clam Sauce, 100
    Pasta Salad with Crab and Snow Peas, 104
    Pasta with Shrimp, Zucchini, and Tomatoes, 103
    Poached Halibut and Peppers, 121
    Scallops in White Wine, 122
    Shrimp Scampi Provincial with Linguine, 107
    Shrimp-Stuffed Artichokes, 123
    Sole Veronique, 124
    Spicy Shrimp, 125
    usage chart, 37
Seafood seasoning, 37
Seasoned salt, 35–38
Sesame seed, 36, 38
Shelflife, herbs, 10. *See also* individual herbs
Shellfish
    about, 114
    Artichokes with Shrimp, 54
    Baked Sole with Shrimp and Crabmeat, 116
    Cajun, 134
    Crab Cakes, 56
    Greek-Style Shrimp, 119
    Linguine with Clam Sauce, 99
    Linguine with Herbed Clam Sauce, 100
    Pasta Salad with Crab and Snow Peas, 104
    Pasta with Shrimp, Zucchini, and Tomatoes, 103
    Scallops in White Wine, 122
    Shrimp Scampi Provincial with Linguine, 107
    Shrimp-Stuffed Artichokes, 123
    Spicy Shrimp, 125
    usage chart, 37
Shortcakes, 247
Shrimp, 37
    Artichokes with Shrimp, 54
    Baked Sole with Shrimp and Crabmeat, 116
    boil, 37
    Dilled Shrimp Salad, 66
    Greek-Style Shrimp, 119
    Pasta with Shrimp, Zucchini, and Tomatoes, 103
    Shrimp Scampi Provincial with Linguine, 107
    Shrimp-Stuffed Artichokes, 123
    Spicy Shrimp, 125
Shrimp Scampi Provincial with Linguine, 107
Shrimp-Stuffed Artichokes, 123

262 Index

Side dishes
    about, 190
    Date Risotto, 196
    Fried Rice, 197
    Greek-Style Orzo, 194
    Indian Raisin Rice, 198
    Lemon Rice, 199
    Mashed Potatoes, 191
    Orzo with Parmesan and Basil, 195
    Oven French Fries, 192
    Potato Pie, 193
    Wild-Brown Rice, 200
Simple Beef Stew, 172
Snow Peas
    Pasta Salad with Crab and Snow Peas, 104
    Snow Peas with Mushrooms, 186
Snow Peas with Mushrooms, 186
Sole, 116, 124
Sole Veronique, 124
Soup
    about, 78
    Corn Soup, 79
    Gazpacho, 80
    Golden Squash Soup, 81
    Mushroom and Barley Soup, 82
    Navy Bean Soup, 91
    Tomato Soup Cake, 249
    Tomato-Herb Bouillon, 83
    Yogurt-Cucumber Soup, 84
Soups
    Chicken Stew, 143
    Lentil Stew, 89
    Simple Beef Stew, 172
    Turkey Chili, 153
Soy-Ginger Marinade, 131
Spanish olives, 44
Spanish paprika, 34
Spearmint, 30
Spiced Apricots, 248
Spices. *See also* Herbs; individual spices
    Allspice, 11, 23
    Anise, 12, 39
    Aniseed, 18
    blends, 18, 23 (*See also* Herb Blends)
    Caraway seed, 15
    Cardomom, 15–16, 23
    Cayenne, 16
    Chart, 35–38
    Chinese Eight-Spice Powder, 18
    Chinese Five-Spice Powder, 18
    Cinnamon, 18, 20–21
    Cloves, 18, 21–22
    Coriander, 23
    Curry, 23
    definition, 3–4
    Dill, 23–24
    Extracts, 38–39
    Fennel, 18, 24–25
    Ginger, 15, 18, 27–28
    history, 3, 4
    Licorice, 18
    Mace, 29, 32
    Mustard, 23, 31–32
    Nutmeg, 29, 32, 35
    Paprika, 33–34
    Saffron, 23
    Star anise, 12, 18
    Szechwan pepper, 18
    Turmeric, 23
Spicy Shrimp, 125
Spinach
    Spinach and Chicken Pâté, 53
    Spinach and Feta, 72
    Spinach and Red Cabbage Salad with Honey-Mustard Dressing, 73
    Spinach, Raspberry, and Walnut Salad, 74
    usage chart, 36
Spinach and Chicken Pâté, 53
Spinach and Feta, 72
Spinach and Red Cabbage Salad with Honey-Mustard Dressing, 73
Spinach, Raspberry, and Walnut Salad, 74
Sponge Cake, 230, 233
Spoon Bread, 218
Spreads
    Black Bean Spread, 88
    Cheddar Cheese Spread, 43
    Chicken Liver Pâté, 51
    Country Pâté without Liver, 52
    Eggplant Caviar, 48
    Feta Cheese Spread, 45
    Goat-Cheese Spread, 46, 57
    Guacamole, 49
    Hummus, 50
    Spinach and Chicken Pâté, 53
    Vegetable Cream Cheese Spread, 47
Squash
    Golden Squash Soup, 81
    Stir-Fried Zucchini and Yellow Squash, 187
    usage chart, 36
Star anise, 12, 18
Starters
    About, 42
    Artichokes with Shrimp, 54
    Black Bean Spread, 88
    Cheddar Cheese Spread, 43
    Cheddar Olive Spread, 44
    Chicken Drumettes, 55
    Chicken Liver Pâté, 51
    Country Pâté without Liver, 52

Crab Cakes, 56
Eggplant Caviar, 48
Feta Cheese Spread, 45
Goat-Cheese Spread, 46, 57
Guacamole, 49
Hummus, 50
Phyllo Goat Cheese Squares, 57
Spinach and Chicken Pâté, 53
Vegetable Cream Cheese Spread, 47
Stews
   Chicken Stew, 143
   Lentil Stew, 89
   Navy Bean Soup, 91
   Simple Beef Stew, 172
   Turkey Chili, 153
Stir-Fried Zucchini and Yellow Squash, 187
Storing herbs, 6–10. *See also* individual herbs
Strawberries, 67
Sun-Dried Tomatoes, Rosemary, and Thyme Pasta, 108
Sweet basil, 12–13
Sweet cumin, 12
Sweet laurel, 13–14
Sweet marjoram, 29–30
Sweet potatoes, 36
Swordfish, 117
Szechwan pepper, 18

T

Tahini, 50
Tarragon, 36, 37
Thailand, flavorings, 4
Thyme
   Herb blend, 40
   Sun-Dried Tomatoes, Rosemary, and Thyme Pasta, 108
   usage chart, 35–38
Tomato Soup Cake, 249
Tomato-Basil Sauce with Mostaccioli, 109
Tomato-Herb Bouillon, 83
Tomatoes, 35
   Basic Marinara Sauce, 95
   Chicken Cacciatore, 136
   Chicken Cakes with Tomato and Sweet Pepper Sauce, 137–138
   Chicken Chasseur, 139
   Colorful Pasta Peppers, 96
   Cucumber and Tomatoes in Lettuce Cups, 65
   Dilled Shrimp Salad, 66
   Fettuccine with Artichokes and Tomatoes, 98
   Gazpacho, 80
   Greek Salad, 68
   Greek-Style Orzo, 194
   Greek-Style Shrimp, 119
   Ground Beef with Rice, 161
   Halibut Cacciatore, 120

Lentil Stew, 89
Linguine with Clam Sauce, 99
Linguine with Herbed Clam Sauce, 100
Meat Sauce, 102
Moussaka, 166–167
Navy Bean and Basil Salad, 90
Pasta with Shrimp, Zucchini, and Tomatoes, 103
Pork with Tomato-Rosemary Sauce, 169
Pot-Roasted Pasta, 105
Prosciutto and Tomato Pasta, 106
Sun-Dried Tomatoes, Rosemary, and Thyme Pasta, 108
Tomato Basil with Fresh Mozzarella and Balsamic Vinegar, 75
Tomato Soup Cake, 249
Tomato-Basil Sauce with Mostaccioli, 109
Tomato-Herb Bouillon, 83
Turkey Chili, 153
Turkey Sausage with Peppers and Onions, 156
Tricolor Peppers, 57, 188
Turkey, 36
   about, 128–129
   Meatloaf, 165
   Turkey Breast Marinated with Lemon and Herbs, 152
   Turkey Chili, 153
   Turkey Sausage, 154, 155, 156
Turkey Breast Marinated with Lemon and Herbs, 152
Turkey Chili, 153
Turkey Sausage, 154, 155, 156
Turmeric, 23

V

Vanilla extract, 39
Veal, 38
Vegetable Herb Blend
   Bibb Lettuce with Radishes, 63
   Cheddar-Beer Bread, 212
   Chicken Stew, 143
   Cucumber and Tomatoes in Lettuce Cups, 65
   Fried Rice, 197
   Goat Cheese Spread, 46
   Herbed Spoon Bread, 218
   Macaroni and Cheese, 101
   making, 40
   Orange and Onion Chicken, 148
   Oven-Fried Chicken, 150
   Savory Biscuits, 222
   Tricolor Peppers, 188
Vegetables. *See also* Salads; individual vegetables
   about, 174
   Artichokes, 175
   Brussels Sprouts, 176
   Carrots, 177
   Chicken Strips with Brussels Sprouts, 144
   Green beans, 178

Vegetables (*Continued*)
    Green Beans Almondine, 179
    Green Beans Vinaigrette, 180
    Julienne Carrots and Zucchini, 181
    Lemon-Herbed Asparagus, 182
    Mushrooms Paprika, 183
    New Potatoes and Carrots, 184
    Oven-Fried Eggplant, 185
    Snow Peas with Mushrooms, 186
    Stir-Fried Zucchini and Yellow Squash, 187
    Tricolor Peppers, 188
Veggie Omelet, 207
Vermicelli
    Basil and Pine Nuts, 110
    Cream Cheese and Herbs, 111
Vinaigrette
    Basic, 69
    Green Beans Vinaigrette, 180
Vinegar
    Balsamic, 75
    Basil Marinade, 133

W

Waffles, 208
Waldorf Salad, 76
Walnuts
    Apple-Walnut Cake, 229
    Apricot Nut Bread, 211
    Chocolate Chip and Walnut Scones, 214
    Cinnamon Nut Bread, 215
    Cinnamon Rolls, 216
    Nut Bread, 219

Pumpkin Cake, 244
Pumpkin Muffins, 221
Romaine, Pears, and Walnuts, 71
Spinach and Feta, 72
Spinach, Raspberry, and Walnut Salad, 74
Tomato Soup Cake, 249
Waldorf Salad, 76
Yogurt-Cucumber Soup, 84
Water chestnuts, 66
Water mint, 30
White pepper, 35, 37
Wild-Brown Rice, 200
Winter squash, 36, 81

Y

Yams, 36
Yellow squash, 36, 187
Yogurt Cheese, 128, 227
Yogurt Cream, 128–129, 228
Yogurt-Cucumber Soup, 84

Z

Zest
    Lemon, 29, 152
    Orange, 33, 148
Zucchini, 36
    Chicken Strips with Brussels Sprouts, 144
    Julienne Carrots and Zucchini, 181
    Pasta with Shrimp, Zucchini, and Tomatoes, 103
    Stir-Fried Zucchini and Yellow Squash, 187